STATISTICAL THINKING

A STRUCTURAL APPROACH

SECOND EDITION

John L. Phillips, Jr.
BOISE STATE UNIVERSITY

W. H. FREEMAN AND COMPANY
SAN FRANCISCO

Project Editor: Patricia Brewer
Copyeditor: Elisabeth M. Magnus
Designer: Nancy Benedict
Cover Designer: Dare Porter
Production Coordinator: Bill Murdock
Illustration Coordinator: Cheryl Nufer
Compositor: Graphic Typesetting Service
Printer and Binder: The Maple-Vail Book
    Manufacturing Group

Library of Congress Cataloging in Publication Data

Phillips, John L., 1923–
    Statistical thinking.

    (A Series of books in psychology)
    Includes index.
    1. Social sciences—Statistical methods. 2. Psy-
chometrics. 3. Statistics. I. Title. II. Series.
HA29.P517   1982   519.5   81-17368
ISBN 0-7167-1379-9   AACR2
ISBN 0-7167-1380-2 (pbk.)

Printed in the United States of America

234567890   MP   0898765432

To Mark and Steve

CONTENTS

PREFACE

The purpose of the second edition of *Statistical Thinking* has not changed from that of the first edition. The book is not intended to be the main text of a statistics course; in such a course it should be assigned *before* work on the main text begins. Thus used, it can be helpful in breaking the psychological barrier that stands between many students and statistics— especially when statistics is a required course. Primarily, however, the book will be used as basic reading in the statistics or measurement sections of courses with broader content domains—courses in the behavioral sciences or in professional curricula (business, education, social work, and so forth) that are related to the social sciences. It can also be used in any course concerned primarily with measurement (for example, Educational Tests and Measurements) in which the instructor does not assume a prior acquaintance with statistics. Any number of chapters can be covered, providing that the chapters assigned are consecutive and begin with Chapter 1.

This book is for beginners, and beginners often have trouble conceptualizing. In teaching statistics to college students with no background in the discipline, concrete images are important. The early chapters, therefore, are liberally laced with such images. Only gradually is the student brought to the second-order abstractions of Chapter 8.

Even then, the student is invited to retreat occasionally—to consolidate learnings from earlier topics and to build confidence for the ones ahead. Many otherwise competent college students are frightened into confusion when confronted by the necessity of thinking quantitatively; they have had unsuccessful experiences with such demands in the past. But much of their travail has been related directly to the manipulative aspects of quantitative thinking—aspects that have been emphasized in their experience at the expense of logical structures. Such students might comprehend quantitative concepts *if manipulations were deemphasized and if concepts were carefully and logically developed.*

If one is training *practitioners* in statistics and measurement, manipulations are important in themselves; but students who will be statistical "consumers" need to understand basic concepts in order to know what the manipulations are about and comprehend the results of manipulations done by others. Even the manipulatory skill of the practitioner must be built upon a solid understanding of the structure that underlies every manipulation. This book therefore functions in two ways. First, it presents basic ideas in the form in which they are needed by consumers of statistics. Second, for the relatively few who are destined for more advanced training in statistics, it provides a helpful framework for the more detailed study that lies ahead.

The metaphor "framework" is apt, because the approach of the book is structural. It is concerned with binding together ideas—relating each new idea to others already familiar to the student or presented earlier in the text. Cross references are frequent and details are subordinated to overarching principles.

Sample Applications

At the suggestion of users of the first edition, I have added sample applications to this second edition. It is true that this book was written for *consumers*, not practitioners. However, a good way to get students to think about possible applications is to ask them to choose the appropriate statistics for given situations and explain why each choice was made. Consumers do not make such decisions, of course; but taking the role of practitioner should deepen student involvement. And since the solutions to these problems do not demand the computational skills of the practitioner, there is not only much to be gained but also little to be lost by trying them.

All of the problem situations in each of five disciplines (education, political science, psychology, social work, and sociology) were suggested by representatives of those disciplines. Each respondent identified, in his own discipline, a legitimate application of each of several key concepts and then explained why each application was appropriate. Edited versions of their descriptions appear at the ends of the chapters in which the concepts are discussed (Chapters 3 through 9); explanations are at the back of the book.

Some Specific Suggestions

Average college students, if they are motivated to succeed and confident that they can, should comprehend most of this text entirely on their own. However, a few ideas are exceptionally difficult, and even a good student might appreciate some help.

Confidence Levels

The section that introduces the concept of confidence levels (Chapter 7) is difficult because the student must (1) imagine a true distribution of means with its own (population) mean in a different location from that of the obtained distribution and then (2) find the probability of obtaining what he or she has indeed obtained. This process is a reversal of ordinary patterns of thought in which established fact constitutes the standard against which other possibilities are judged. Because the confidence-level operation is such a strange mental maneuver, I have provided some suggestions for practicing it. The instructor can help by assigning the exercises and requesting answers (that is, probabilities acquired by inspection) and then displaying the same operations via overhead projector, using a translucent "distribution cutout" so that the vertical line above the obtained mean remains visible as the distribution is moved about on the base line.

The z Ratio

The section on the z ratio is difficult because there are so many steps in its logical development, each one further than the last from the concrete empirical fact of test performance. Here the instructor can help by tracing the steps carefully, beginning with the selection of two sets of sample means from a population (of French grammar students, all of whom have had the same quality of instruction), and continuing through all the diagrams in Figures

8-1 through 8-3 to the placement of the obtained difference within a distribution of differences in Figure 8-4. The population diagram can be drawn on a chalkboard to emphasize its size relative to the others, and Figures 8-1 through 8-3 can be reproduced on a transparency for overhead projection. In any case, what the instructor adds to the text is *movement*—an emphasis on the transformations that occur from one diagram to the next.

Analysis of Variance

In many courses, assignment of the section on analysis of variance will not be appropriate. Students who do progress that far will find that the central problem is the extraction of *variance among means* from *total* variance. It may be helpful to demonstrate the component character of each individual score by analyzing several such scores in the context of Figure 9-1. The instructor may choose any point on the base line as a score to be analyzed, designate that score as a part of one of the six distributions, and then show how the deviation of that score from the grand mean can be described as the algebraic sum of (1) its deviation from the mean of its own distribution and (2) the deviation of that mean from the grand mean. The process may be repeated for any value on the scale.

While Figure 9-1 is still on display, the instructor can show with a gesture how the elimination of variability among means would result in a collapsing of all six distributions into the single distribution portrayed in Figure 9-2. Such a preview of the section on analysis of variance (with ensuing discussion) may save students both time and frustration. Again, what a text cannot do is to *move,* and transitions are best communicated by movement.

Acknowledgments

I am still grateful to all of the persons whose assistance with the first edition is acknowledged there. Two of them, Mark Snow and Steve Thurber, of the Department of Psychology of Boise State University, have been recognized in the dedication of this book.

Contributions to the second edition were made in the form of sample applications in five related fields of endeavor. The contributors are all colleagues at Boise State University:

Education	Dr. Keith Brownsmith
Political science	Dr. Greg Raymond
Psychology	Dr. Steven Thurber
Sociology	Dr. James Christensen
Social work	Mr. David Johnson

January 1982   John L. Phillips, Jr.

STATISTICAL
THINKING

1 | INTRODUCTION

You are probably studying statistics not because you want to but because you have to. If so, I know how you feel. I went through the same experience years ago; if I could have avoided statistics, I probably would have. However, my attitude changed after I began to study it, for I discovered in it a new way of thinking that really fascinated me.

But your task is in an important way more difficult than mine was. You won't have to do the computations that I did, but you are being asked to acquire in a fraction of one semester (or quarter) an understanding of the same basic concepts to which I devoted two full semesters.

The Task

Your plight and your prospects are well illustrated, I think, by the experience of a student who was asked to evaluate the prototype of this book:

> It was the most difficult book I have ever read. It was foreign to me since I had had no background knowledge of the things talked about. . . . If I was to understand it, I realized I would need to outline the book chapter by chapter. I did so, and to my amazement, it followed a very orderly pattern after all. It really did present what the author had stated he hoped to present. If you understood Chapter 1, you could see the logic of Chapter 2, and so on through each chapter. I feel I learned a great deal about measures in a relatively short period of time.

That quotation also contains some important advice on how to use the book. I would only add that even though you may have mastered the ideas preceding the one you are working on at any given moment, you should be prepared to go back to those ideas from time to time and consider their relation to the new one being presented. I have tried, by providing frequent cross references, to help you do just that. (You may wish to keep a couple of bookmarks handy for this purpose.) When you are finished, you should have in mind a hierarchical *structure*, with each new idea related to one or more of those that have preceded it.

The acquisition of such a structure is highly satisfying in itself, but that probably is not the reason that a unit on statistics has been included in the course you are now taking. Why, then, has it been included? An understanding of statistical concepts will not be of critical assistance to you in gathering economic data, in conducting interviews for a political or sociological opinion or attitude study, in uncovering archeological artifacts, or in teaching children. But often people who do economic, political, sociological, anthropological, archeological, or educational studies (to name but a few) report their findings in statistical terms. If you are planning a career in any of the several professions to which those studies are relevant, it is important that you be able to read them with understanding. A continuing acquaintance with developments in his or her field is the mark of the true professional, and this book will help you maintain that acquaintance.

The Basic Ideas

To understand the meaning of any measurement in the social sciences, you must come to know at least two things about it. First, you must be able to describe the operations by which it was obtained, and second, you must be able to place it in relation to other measurements that have been obtained in the same way. You may also find it helpful to place it within some theoretical framework.

This book is concerned primarily with the second kind of knowledge. Statistical thinking deals with multiple measurements. It analyzes the relation of "how many?" to "how much?"— of frequencies to scores. If the basic element in measurement itself is a *score*, the corresponding concept in statistics is a *distribution* that includes many scores—in short, a *frequency distribution*.

Such a distribution can be described by drawing a picture of it—and indeed in the early stages of your learning about distributions, that is the method I shall use to describe them. The method is cumbersome, however; so ways have been devised to achieve roughly the same result through the use of numbers rather than diagrams. The most important advantage of numbers over diagrams is that they can be manipulated in a way that diagrams cannot.

Imagine that (for reasons known only to your psychoanalyst) you have just had a pile of gravel dumped on your front lawn and that (for similar reasons) you want to describe the result to me over the telephone. To do that successfully, you will have to tell me at least three things about the pile: (1) its general configuration—that is, whether it is shaped like a cone, a pancake, or your garage roof, (2) its location—that is, how far and in what direction it is from some reference point familiar to me, and (3) the extent to which it is "spread out"—for example, if it is a cone, is it a steep-sided one that covers only a small area, or is it a low-profile cone that covers most of the lawn?

That pile of pebbles is analogous to a frequency distribution of scores, and the same kinds of information are needed to describe either adequately. Concerning configuration, we have certain names for frequency distributions that convey information to anyone who is familiar with them—names such as "normal," "symmetrical," "positively skewed," and "bimodal." Concerning location, the procedure is pretty much the same as it is for a pile of gravel. A dimension is identified and a reference point is chosen; measurement of distance to the center of the distribution is from that reference point. The result is a *measure of central tendency*. And finally, to communicate information about the amount of dispersion, a new reference point is used—namely, the center of the distribution—and the needed information may consist of the average distance of individuals (like that of the individual pebbles in a pile of gravel) from the central point. That distance is a *measure of variability*.

But just describing a distribution is not always enough. Often you will be interested in two distributions and in the relationship that exists between them. Consider a single variable, "IQ in the general population," and a few other variables with which it might be paired: family income, some index of health care, a general index of socioeconomic status, race, or place of residence. Other interesting relationships could be investigated among the IQs of various subgroups of the general population: between parents and their children, between identical twins, between fraternal twins, between non-twin siblings, and between pairs of unrelated children. These are just a few that come to mind at the moment. Others would occur to you if you were making a study of intelligence and its correlates, and in every case you would need a way of communicating your findings to others; in short, you would need a *measure of relationship*.

Whether or not you wish to relate one set of scores to another, you will surely want to be able to tell what each individual measure means. If you are a teacher who has given my child a test, and I ask you how well he did on it, you might try to put me off with a "high" or a "low" and go on to talk about something else. But if I want to know "How high?" you are in trouble. You may answer that he got 90 percent of the test items right. You think you are off the hook, but I persist: "How hard is that test? 'Ninety percent' is very impressive if the items are all difficult, but not if most of the other kids scored above ninety-five!"

Our conversation has been concerned entirely with the *interpretation of individual measures*, and I'm sure you'll agree that all of my questions are pertinent. Without answers to them and others like them, one really cannot know the meaning of a score.

On the other hand, you must be careful to avoid overinterpreting measures, whether they be of individuals or of groups. If, for example, you were to weigh a *random sample*[1,2] of fifty ten-

---

1. A random sample is one in which (a) every member of the population has an equal chance of being included in the sample and (b) each selection is made independently of all the others.

2. The notes in this book are of two kinds and are denoted by two kinds of superscript. Footnotes are indicated by numbers([1], [2], [3], etc.); other notes, which appear in the back of the book, are indicated by asterisks (*).

    The footnotes are intended to be supplementary to the central discussion at any given time and are important to a full understanding of that discussion. A footnote may provide (a) refinement of an idea by restricting or extending its implication, (b) explanation of a pedagogical technique by commenting on the relative importance of its various attributes or (c) cross references that are not essential but are enriching. In fact, the objective in every case is enrichment. The way to use the footnotes, then, is (1) *to ignore them* as you work through a section or chapter for the first time and (2) *to study them carefully* the second time through. Every chapter has a few footnotes.

    The notes in the back of the book are more technical and go beyond the scope of the text. Ignore them at least until you have mastered the chapter to which they refer. These notes pertain to subjects mentioned in Chapters 3, 4, 5, 8, and 9.

year-old boys, could you easily compute a measure of central tendency for the *sample* that is identical to the central tendency of a population that includes *all* ten-year-old boys? How different might the obtained average be if it were computed from a different random sample of that same population? Questions like these have to do with *precision of measurement,* and we do have ways of dealing with them.

Assuming for the moment that you do know how to deal with questions about precision, consider this one additional question: Upon further analysis of your sample of ten-year-old boys, you discover a marked difference between the weights of boys living in one area of the country and those living in another. You suspect that the difference is due to diet, and you subsequently narrow that hypothesis down to a single vitamin that seems to be more plentiful in one of the areas than in the other. One way to test your hypothesis would be to select two samples of male infants who are living in the area in which the vitamin is least plentiful, introduce the suspected vitamin into the diet of one group, and then after a period of nine years weigh them both again. Let's imagine that there is a difference between the two. Is it large enough that you can be reasonably sure that it did not occur by chance—that a replication of the same study would not turn up a difference of zero, or even a difference in the other direction? To put it another way, "How *significant* is the difference you obtained?" Again, there are ways of dealing with such questions.

In the chapters that follow, each of the above ideas will be developed further, but always in the manner that you have seen here. If you are a sophisticated mathematician—or a frustrated accountant—this book is not for you; it is aimed directly at the underlying *logic* of statistical thinking, with an absolute minimum of arithmetical and algebraic manipulations. You will find the logic similar in many ways to common sense. The main difference is that it is rigorously systematic, and like any system, its parts are interdependent. So the book cannot be studied piecemeal; the ordering of the chapters was deliberate and necessary. Once you have worked through it, however, the book will serve as a convenient reference to take with you into your professional life. It has been organized with that in mind.

2 | FREQUENCY DISTRIBUTIONS

Every year, Cerebral University administers a scholastic aptitude test to each of its incoming freshmen. After the tests have all been scored, each freshman is handed a small card with his score on it; then all the freshmen are herded into the seating area of the football stadium.

A university official stands on the sideline of the football field with a microphone in his hand. He points to the space between the west goal and the adjacent five-yard line and announces that anyone with a score below 5 should come down and stand in the middle of that space. Nobody moves, so he walks five yards to the east and repeats the instructions for that space. There is a long pause, then one miserable soul slinks down to the field; he is the only one in the *class interval* 5 through 9.[1] Another call (10 through 14) yields two students, a fourth (15 through 19) gets him five, a fifth (20 through 24) 13, and so on, with increasing *frequencies* up through the middle scores, followed by decreasing ones after that. Figure 2-1 is an aerial view of the field after all of the students have left their seats.

---

1. Technically, "5 through 9" is only an *apparent* class interval. Its precise limits—often called "true limits"—are 4.5 and 9.5. Those of the next higher interval are 9.5 and 14.5, and so on up the scale. This adjustment of the class intervals is required because test scores are given by discrete or unique integers such as 5, 6, 7, 8, and 9, while the precise limits refer to the continuous intervals extending from 4.5 to 5.5, 5.5 to 6.5, 6.5 to 7.5, 7.5 to 8.5, and 8.5 to 9.5. The continuous interval 4.5 to 9.5 is the sum of all those.

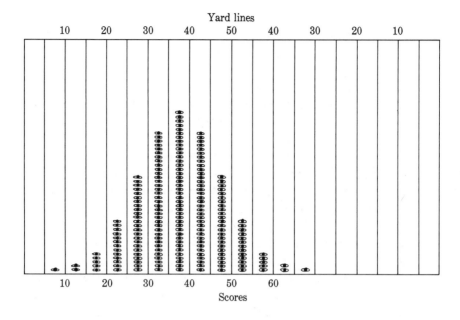

Figure 2-1 Students standing on football field in columns designated by aptitude test scores.

Normal Distributions

Cerebral U. lacks the budget for it, but if our official had a rope long enough to run from the sideline, out around the entire freshman class, and back to the sideline (and if the official were indeed to use it in that way) the rope would form what is known as a *normal curve.*[2] It would be so called because it would enclose a *normal distribution.*

Many quantities that are complexly determined do tend to form normal distributions. Scholastic aptitude measures often show that tendency. Such measures have many determiners, both hereditary and environmental; some of those determiners influ-

---

2. It is sometimes called a bell curve, because of its bell-like shape. If instead of draping the rope loosely in a smooth curve, our official were to instruct the end student in each column to grasp it firmly and pull it taut, the rope would form a series of straight lines and angles known as a *frequency polygon.*

ence an individual's score in an upward, some in a downward direction. In a few individuals, there is a preponderance of positive determiners; theirs are the scores in the upper (right-hand) *tail* of the distribution. In a few others, negative determiners predominate; their scores form the lower tail. Most scores, however, represent more balanced combinations of determiners; they form the large middle area of the distribution.

All of this is assumed to be a matter of chance, and you can readily see that the probability of, say, 100 determiners being all positive (or all negative) in any individual would be vastly smaller than the probability of their being approximately half and half. If you don't believe that, toss a mere three coins[3] just eight times[4] and plot the number of occurrences against the numbers of heads that could possibly occur in a single toss (0, 1, 2, and 3 are all possibilities). You should come up with a distribution similar to the one in Figure 2-2. "All heads" (score = 3) does not occur often by chance, and neither does "all tails" (score = 0). The middle scores are much more frequent.

By the way, do you get the impression when looking at Figure 2-2 that if there were more class intervals and a larger number of observations the distribution of coin tosses would look very much like the approximately normal distribution of students that we got from Cerebral U.? If you do, you are right; the larger that number is, the more closely the actual distribution approximates the mathematical model we call "the normal curve." (This is the first of several references to effects of sample size. The symbol for the number in a sample is *N*; a *sample* is made up of the individuals or observations—that is, the number of *scores* under discussion. In the case of the CU freshmen, *N* is 200; in that of the coin tosses, it is 8. The relation of the sample to its *population* is discussed on pages 4 and 15–16.)

A frequency distribution can deviate from normality in several ways, but we shall mention only four of them here: positive skew, negative skew, J, and bimodal.

---

3. Three coins correspond to three determiners instead of the 100 cited in the example we just considered.

4. Eight tosses correspond to eight students as compared to the 200 CU freshmen.

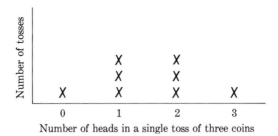

Number of heads in a single toss of three coins

Figure 2-2 Distribution of eight coin tosses.

Skewed Distributions

If the lower tail of a distribution is extended abnormally, that distribution is said to be skewed, and since the direction of the extension is downward it is further categorized as *negatively* skewed. If your instructor in this course should give you a very easy test, the distribution of your scores might look rather like Figure 2-3, which illustrates a negative skew.

If the upper tail is similarly extended, the skew is said to be positive, and the distribution is *positively* skewed. A difficult test would produce a distribution more like Figure 2-4, because only a talented (or industrious) few deviate very far from the lowest possible score.

Scores

Figure 2-3 Negatively skewed distribution.

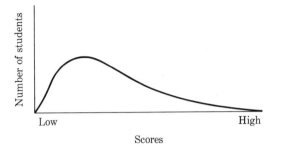

Figure 2-4   Positively skewed distribution.

Other Configurations

There are other possibilities. If we were measuring conformity behavior, like that of motorists at a busy intersection, we might get a J *curve*, like the one in Figure 2-5. The same would be true of a distribution of scores on a test that is extremely easy (so easy that most people get perfect scores) or extremely difficult (so difficult that most of the scores are zero).

An entirely different configuration would emerge if we were to measure the standing height of humans; for there are two phys-

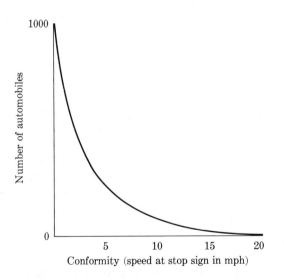

Figure 2-5   J curve of conforming behavior.

ical types of humans—male and female—and that should result in a bimodal distribution, as shown in Figure 2-6. Or, if your instructor were to spring a pop quiz on Chapter 5 of this book at a time when only half of the class had read it, the distribution of those scores would be bimodal. (Note that in Figure 2-6 the distribution is apparently quite symmetrical; but a bimodal distribution can be asymmetrical, as indeed this one surely would be if, say, it were composed of twice as many women as men.)

But the J distribution is difficult to deal with statistically, and a bimodal distribution can be dealt with by separating the two normal distributions that are partially concealed within it. So these "other configurations" are of only passing interest to us here.

Summary

Many frequency distributions in the social sciences are approximately "normal"; that is, in the typical distribution there are a few very low scores and a few very high ones, but the great mass of individuals tends to pile up on the middle scores. This occurs because the probability of all the many determiners of a trait pointing in the same direction is virtually nil, and that of a balanced combination of determiners is much higher. Although other configurations do occur (positive skew, negative skew, J curve, and bimodality are mentioned), our primary concern in this book will be the normal distribution, because it serves as a good approximation to the kinds of distribution most frequently encountered in behavioral investigations and because it is the configuration for which the best known statistical treatments are available. We shall encounter some of those in subsequent chapters.

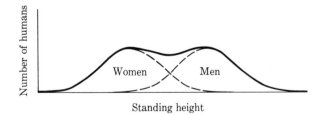

Figure 2-6   Bimodal distribution of humans on a scale of height.

3 | MEASURES OF CENTRAL TENDENCY

Somewhere in the intermediate grades you were introduced to the concept of "average," and you have used it ever since on the assumption that there is only one such. Actually, there are several averages, of which three will be described here.

An average can represent either the sample itself or the *population* from which the sample is taken. A population comprises all the individuals who share certain specified characteristics. We might speak of the average scholastic aptitude score made by entering freshmen at Cerebral University, for example. If we actually have scores from only those freshmen who entered the University in a particular year—say 1980—our statement of the average for "entering freshmen" is based upon the assumption that the 1980 average is the same as the average for all freshmen who enter the University in all years, past and present.[1] On the other hand, we may use the average to represent only the class whose scores are on hand, in which case our sample average is precisely equal to the population average. Or we may use it to represent entering freshmen everywhere, in which case the sample average probably will not be equal to the population average. We proceed in each case as if our *obtained* average from the sample is also the *population* average. But that assumption is valid only to the extent that the sample really "represents" the population. In other words,

---

1. Actually, of course, CU officials do not herd their freshmen onto the football field. They don't even use data from the current class. What they really do is to compare each new freshman's scores to those of last year's freshmen, on the assumption that there is no significant change in the entering freshman class from one year to the next.

the more we attempt to generalize from an obtained average (or any other statistic, for that matter), the less confident we can be of the accuracy of our statement.

But in science, generalization is the name of the game. There may well be some immediate practical reason for ascertaining the average scholastic aptitude score of freshmen entering Cerebral University in the current year (in that case, the sample would be identical to the population); but when a scientist observes a phenomenon in his laboratory, nothing of importance has been accomplished unless the result can be generalized. The phenomenon is a mere curiosity unless it can be replicated (for example, with subjects of a different species—or at least with other members of the same species). All science is based upon generalization; we almost never study all the individuals of a population in which we are interested. Instead, we study a sample, and one very useful description of a sample is its central tendency or average.

The Mean $(\overline{X})$

The average that you learned about in grade school was a *mean*. The mean is not, as you were led to believe, *the* average; but it does have characteristics that make it the best one to use in many circumstances. Whenever the frequency distribution is fairly symmetrical and plenty of computation time is available, the mean is the statistic of choice. The computation time is necessary because all of the scores must be added together before the sum can be divided by N (number of scores), and there are sometimes thousands of scores to add. The formula for the mean is

$$\overline{X} = \frac{\Sigma X}{N} \qquad \text{(3-1)}$$

where $\overline{X}$ is the mean, Σ is a combining term meaning *sum of*, and X refers to the *raw scores*[2] that have been recorded. The expression ΣX (read "summation ex"), therefore, is the sum of all the raw scores in the distribution. Divide it by N and you have your mean.

---

2. A raw score is one that does not imply a comparison with any other score; "inches," "points," "runs," "hits," "errors," "points," and "number right," when those units are simply counted, are raw scores. Raw scores are the only kind we have dealt with so far; others will be introduced later.

I have given all this attention to the computation of the mean mainly because it gives me an opportunity to introduce you to a few of the symbols that we will need later. Actually, learning to compute a mean should receive very little of our attention for two reasons, the first of which is that you already know how to do it.

The second reason is that computation is not what this book is about; our objective is rather to acquire an understanding of statistical concepts. The concept of the mean can best be understood through the following illustration.

Essence of the Concept

Try to imagine a long beam made of an exotic metal that is (1) totally rigid and (2) utterly weightless. Upon that beam we shall place 14 cubes (to represent 14 scores), each of which has a weight equal to that of each of the others. Figure 3-1 shows one possible distribution of cubes.*[3]

At what point on the beam would a fulcrum (support) have to be placed to establish an equilibrium. That is, what is the *balance point* of the distribution?

This distribution is symmetrical, so it should be easy to see that it would balance if the fulcrum were placed at 5.5. But that is what all the computation is about; by using the formula, we can find the precise point we are seeking[4] even in cases in which a diagram would be rather puzzling. That such situations do arise will become clear in our section on the median.

In any event, a good verbal definition of the mean is "the *balance point* of the distribution."

Appropriate Applications

The most important advantages of using the mean are that it (1) is the most stable measure of central tendency and (2) lends itself to the computation of other notably stable measures.

---

3. Asterisks (*) refer to notes that appear at the back of the book and need not be read. They are there to provide an option for exceptionally motivated students but presumably will be omitted by most. Such a note may be more technical than is necessary for a clear view of the *structure* of a concept; it may introduce a term not used elsewhere or a related fact that is not explained in the text; or it may cite an unresolved controversy among statisticians. In any case, you may skip it with impunity.

4. $\bar{X} = \dfrac{\Sigma X}{N} = \dfrac{\text{Sum of the raw scores}}{\text{Number of raw scores}} = \dfrac{77}{14} = 5.5.$

Figure 3-1   Fourteen cubes on a weightless beam.

By "most stable" I mean that it varies least among repeated random samples selected from the same population. Let us say we are interested in finding the mean Wechsler Adult Intelligence Scale (WAIS) IQ of second-generation Japanese-American (Nisei[5]). For simplicity, let us assume that the population itself remains stable during the time that it takes us to sample it. Such a population would probably be distributed normally; let's assume that it is and that Figure 3-2 is an accurate rendering of that distribution.

Theoretically, we could simply test every Nisei in the country, sum their scores, and divide by their N. Practically, however, that would be virtually impossible; as mentioned earlier, measurements are almost never taken of whole populations. What we actually do is to measure a *sample* of whatever population we wish to study. There are more ways than one of selecting a sample. Conceptually the simplest of them is *random* selection, in which the composition of the resulting sample is entirely a matter of chance. (See footnote 1 on page 4.)

Figure 3-2   Distribution of Nisei on an IQ scale.

5. Pronounced "Nee'say."

Now back to the stability of the mean. If we were to select, say, 1000 individuals at random from the population of Nisei, we could compute a mean of that sample. If we were to return that 1000 to the population and then select another sample of the same number in the same manner, we could compute a second mean. Returning those subjects to the population, we could select a third sample, and a fourth, and so on *ad infinitum*. Each sample would have a mean, and not all of those means would be the same. In fact, they would form a distribution—a distribution having the same bell shape as that of the population and of each of the samples.* It would, however, be a much more compact distribution than any distribution of individuals. Figure 3-3 superimposes distributions of the population, of a single sample, and of the means of many samples.

In Figure 3-3, the mean of the means ($\mu_{\bar{x}}$ or "mu-sub-ex-bar") has been placed at the mean of the population (μ_x, or "mu-sub-ex"), and indeed if an infinite number of means were taken, their mean would be the same as the population mean. But, of course, since that is never done, we can never be sure precisely where the population mean really is. (Remember, we know the population only through the samples that we draw from it.)

What we *can* do is to confine the population mean within certain limits; or more accurately, we can discover the limits within which the population mean probably lies. The horizontally compact configuration in Figure 3-3 is the distribution of sample means; when it is extremely narrow, we know that the population mean is very near to our sample mean.

At this point, let me clear up a possible misunderstanding. We do not actually take one sample after another and plot a distribution of their means; rather, we estimate mathematically[6] what would occur if we *were* to do that. Actually, we have only one sample—hence the reference in the preceding paragraph to "our sample mean"—and every bit of knowledge we acquire about the population comes necessarily from that sample.

---

6. The information that is needed from the sample in order to make this estimate concerns its size and its *variability*. The variability of the sample is discussed in Chapter 4, Measures of Variability, while the estimate itself can be found in Chapter 7, Precision of Measurement.

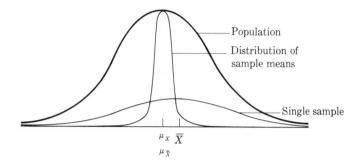

Figure 3-3 Superimposed distributions of population, sample, and sample means.

μ_x = Mean of the population (hypothetical). (A parameter.)

$\overline{X}$ = Mean of one sample (obtained). (A statistic.)

$\mu_{\overline{x}}$ = Mean of an infinite number of sample means (hypothetical)—the same as the mean of the population.

We call any measurement that we make of the sample a *statistic*; the mean IQ of our sample of Nisei is one such statistic. The mean IQ of the population (all Nisei) is called a *parameter* of that population; any other characteristic of the population would be another parameter. The parameters are there whether or not we compute the statistics. In other words, *every statistic we obtain from the sample is an estimate of a particular population characteristic* (parameter).[7]

All this is background essential to the understanding of the proposition that of all known measures of central tendency, the mean is the most stable—that it "varies least from sample to sample of a given population." If we were to take repeated samples of Nisei and compute any other single measure of central tendency for each sample, the distribution formed by those averages would be wider (less compact) than would a distribution of means. The mean varies less from sample to sample than does any other measure of central tendency. That is what is meant when we say it is more stable.

The other significant characteristic of the mean is that it lends itself to the computation of other important statistics.

---

7. See also the section titled "Standard Errors" in Chapter 7.

Among them are the *standard deviation* (see Chapter 4, Measures of Variability), the *product-moment coefficient of correlation* (Chapter 5, Measures of Relationship), and all of the many *standard errors* (Chapter 7, Precision of Measurement). None of them can be deduced without first computing a mean.

The Median (Mdn)

Another way of indicating central tendency is to tell what point on the base line divides the distribution into two equal parts. Note that I said it divides the *distribution* in half, not the *base line*. Look back at Figure 3-1. There, the beam is the base line, and $35/2 = 17.5$ is not the median. Nor is the median defined as a point halfway between the lowest point (1.5) and the highest (9.5), although because of the perfect symmetry of the distribution, it happens to be there in that particular case.

Essence of the Concept

In every case, the *median* is the point between the lower and upper halves of the distribution. (Remember, the distribution is the group of individuals or scores comprising the sample.) In Figure 3-1, that point is 5.5, because there are 7 scores below and 7 above it. In *any* sample with an N of 14, the median will be midway between the 7th and 8th scores (counting up from the bottom of the distribution); in any distribution of 1000 individuals, the median will be the point midway between the 500th and 501st; and so on with as many illustrations as you care to cite.*

Appropriate Applications

"Who is in what half of the distribution?" If that is our question, the answer will depend upon our first finding the median. A more important characteristic of the median, however, is that although it is not quite as stable as the mean, it can be used in situations where the mean would be inappropriate—namely, where the distribution is markedly skewed (page 9); where it is generally symmetrical but includes a few extreme scores at one end; where one end of the scale of scores is not long enough to

1 2 3 4 5 6 7 8 9 10   34 35

Figure 3-4   Fourteen cubes on a weightless beam.

reveal the true variability in the sample,[8] or where any combination of those conditions exists. It may not be too much of an oversimplification to say that the median should be used in preference to the mean whenever the shape of distribution departs radically from perfect *symmetry*. Consider the situation depicted in Figure 3-1. There, the distribution is nearly normal and perfectly symmetrical, so that the mean and the median are at precisely the same place (5.5). Now look at Figure 3-4 to see what happens when that symmetry is disturbed.

In Figure 3-4, we have exactly the same distribution as the one in Figure 3-1, except that two of the scores have been shifted far to the right. The balance point—that is, the mean—has shifted also; it is now 9.5, which, lying as it does above 12 of the 14 scores, is probably an inappropriate index of the central tendency of this distribution.

But what has happened to the *median* as a result of that shift of two scores? It hasn't moved at all! (Count the scores above and below it, and see for yourself.) You may say that it should have moved—at least a little bit—because the distribution is different than it was before; but, even though it is insensitive to that change, I'm sure you will agree that in this distribution the median is a better measure of central tendency than the mean, because the two extreme scores do have too much influence on the mean. Often, what has happened in such cases is that scores have been forced into one category when they should have been classified into two or more. For example, if you were to plot a distribution

---

8. An additional principle applies here. Because some of the scores are indeterminate, there is no way to know "how far out on the bar" the weight should be placed, and it is the placement of such weights that determines where the balance point is.

of the annual incomes of a football coaching staff, you would probably find that most of them are rather close together but that the
salary of the head coach is distinctly separated from the others. If
you had to report a single average of coaching salaries at Cerebral
University, would you use the mean or the median? (Incidentally,
if you ever want to compute a median, consult any standard statistics text about what to do when the median is *within* a class
interval instead of—as in our example—*between* two class intervals, and within a few minutes you'll be ready to proceed.)

And always remember that any statistic (in this case the
median) is a *measurement* of a sample but only an *estimate* of the
corresponding parameter.

The Mode

The last of the three averages to be presented here is also the
easiest. The *mode* is simply the point at which the greatest frequency occurs.* In Figure 2-1, the mode is 37; in Figure 3-1, it is
5.5; in Figure 3-4, it is 5.0.

The mode's very simplicity explains one of its two main uses:
it is used when a very *quick* estimate is needed. It is also used for
the specific purpose of identifying the most common score.

With the exception of those two special situations, the mode
probably should not be used at all, for it is more subject to chance
fluctuations (it is less *stable*) than either the mean or the median,
and it is the one measure of central tendency that may have more
than one true value.

Summary

Statistics are numbers that carry information pertaining
directly to *samples* and indirectly to *populations*. Unless an entire
population is sampled, the sample may differ from the population;
and since it is the latter in which we are interested, it is important
to know which statistic is least likely to deviate significantly from
the one we would obtain if we were to measure the entire population.

That statement is true of statistics of all kinds. If we are
interested in "measures of central tendency," the *mean* is the statistic that meets our requirements—it is the one that varies least

from sample to sample of any given population. The mean is also a companion to other stable statistics; in fact, it is a prerequisite to the computation of many of them. When the distribution is symmetrical and there is sufficient time for its calculation, the mean is the preferred measure of central tendency.

The *median* is less stable than the mean (but more so than the mode). It is used in the few cases in which the information wanted is specifically the point at which a distribution can be cut in half, but probably its major virtue is that it can be used with distributions that are not symmetrical. Unless the sample is very small, the median is also easier to compute than the mean—that is, the computation requires less time.

Finally, the *mode* takes the least time to compute and is the least stable. It should be used only as a preliminary estimate or as an index of "the most common score" or "the typical case."

In a skewed distribution, the arrangement (order) of the three measures along the base line is predictable. If the skew is negative, as in Figure 3-5, the arrangement is, from left to right: mean, median, mode. If it is positive, as in Figure 3-6, the order is just the opposite: mode, median, mean. Conversely, if you know the order of the three averages, you can tell the direction of the skew.

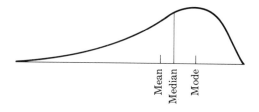

Figure 3-5   Negatively skewed distribution.

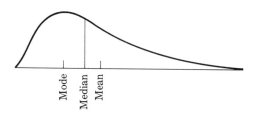

Figure 3-6   Positively skewed distribution.

SAMPLE APPLICATIONS
FOR CHAPTER 3

The following problems require that you make some decisions about the proper use of statistics. In each case, you are presented with a situation that might be faced by a practitioner of the indicated discipline. Imagine yourself as that person in that situation.

Make a tentative choice without looking back into the chapter; explain it as best you can. Then review the text and refine your answer. Finally, check the back of the book (pages 153–174) for a suggested choice and a brief discussion of it.

You may use that choice and that discussion to correct or further refine your response. But if you feel that yours is as good as ours, discuss it with your instructor. If he agrees, I'd very much like to see what you have done.

EDUCATION

A cooperative vocational education program has recently opened to provide one-year training programs for 2000 twelfth grade students from ten school districts. The teachers are in the process of selecting and developing curriculum materials, but they are unsure of the appropriate reading level for the materials. They decide to get an estimate of the reading skills that typify the students enrolled in the program, and you are called in as consultant. Since they have funds to test only 500 students, you select a random sample of 500 from the 2000 who enrolled for the current year. The teachers administer a standardized reading comprehension test and obtain for each student a single score which indicates the reading level of that student. Now what do you do?

POLITICAL SCIENCE

You are studying the domestic and military expenditures of European nations. You want to find the average amount spent by European countries on arms. You cannot afford to study all European countries, so you take a random sample. Now what do you do?

PSYCHOLOGY

You are a family counselor working with the mother of a three-day-old infant. The mother is very concerned about her child (her sister has recently given birth to a "defective" infant) and asks you if her infant is showing normal behaviors for a newborn. You observe the infant in question, but you are not certain what behaviors would be classified as "normal" for a neonate. Your task, therefore, is to find out how newborns tend to behave. You go to three hospitals in your city, visit the neonatal units, and observe and measure infant behavior. For example, you dangle a large red ring in front of each child to see whether he follows it visually or attempts to grasp it. You sound a bell close to each infant's right ear and observe whether the infant turns toward the sound. You then assign points based on your observations (e.g., one point for following the ring visually; two points for grasping the ring). What single statistic would best represent all the children you have tested?

SOCIAL WORK

A child welfare agency director is interested in the length of time that families receive protective services. She asks you to provide information regarding the number of treatment hours received by these clients. To approach this problem you select a random sample of clients from the closed case files. How can these data be analyzed?

SOCIOLOGY

A city council of a city of 100,000 wants to know the average income of its residents. You are asked to make an estimate but given a small budget to do it. You draw a sample of 100 residents and ascertain the average income of each resident. What computation is appropriate?

4 | MEASURES OF VARIABILITY

Different populations (and the samples extracted from them) have different central tendencies, but they differ in another significant respect as well. Consider the two curves depicted in Figure 4-1. Both represent distributions of the same area (identical N's), and both have the same central tendency; nevertheless, the two distributions are very different. In what way are they different? You can see that one is "spread out" more than the other. Since the base line on which the spreading occurs is a single scale of scores, the spreading means that the scores in that distribution *vary* more than those in the "squeezed together" distribution. There are many occasions (some of which are discussed in Chapter 6, Interpreting Test Scores) when it is important to have some kind of numerical index of the variability of a set of scores.[1] In this chapter, we shall examine three such indices.

The Standard Deviation (*S*)

Taking its importance on faith for the moment, let us consider how an index of variability might be devised if we had none already available. It may help to have a concrete example in mind during the discussion, so let us imagine that the scholastic aptitude test mentioned in Chapter 2 is actually a *pair* of tests—one

---

1. Even if we had a diagram of every distribution drawn to scale so that we could compare variabilities by inspection, there would still be a need (demonstrated in Chapter 6) for an index that can enter into mathematical operations that are essentially numerical; you can't multiply or divide a visual perception with an acceptable degree of precision.

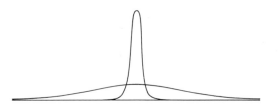

Figure 4-1 Two distributions with the same N's but different variabilities.

of verbal, the other of numerical, aptitude. Their distributions might look something like Figure 4-2.[2] In that diagram, the means of the two distributions have been aligned so that we may concentrate on their respective variabilities. As we look at the figure, we are impressed immediately by the striking difference between the dispersions of the two distributions across their base line. But it is not enough to be impressed; we need a numerical index of dispersion (variability).

Essence of the Concept

How might such an index be devised? One way would be to compare every score in the sample to every other and to average the differences obtained thereby. But that would be entirely too cumbersome, especially with large distributions; and we can get the same effect by selecting a point of reference in the middle of the distribution and measuring the distance of each individual

---

2. The indicated central tendencies and variabilities of the two part-score distributions are such as to make somewhat plausible the proposition that if combined, they would form the distribution shown in Figure 2-1. However, that was not the primary consideration in their selection. Rather it was *simplicity*. We are deemphasizing computation; therefore, computations that *are* required have been made as easy as possible. It is easier to think about an interval that extends from 20 to 25, for example, than one that extends from 23 to 27, or even from 23 to 28. You will find that you can do most—possibly all—of this book's computations entirely in your head.

　But do not attempt any computations unless the requisite data are readily available. For example, in Figure 4-2 do not attempt to confirm the standard deviations of 15 and 5 in the two distributions. To do so, you would need a list of the individual scores, and those have been withheld deliberately to encourage you to focus on the big picture—the configuration of the entire sample for each test and the comparison of two configurations that are very different from each other.

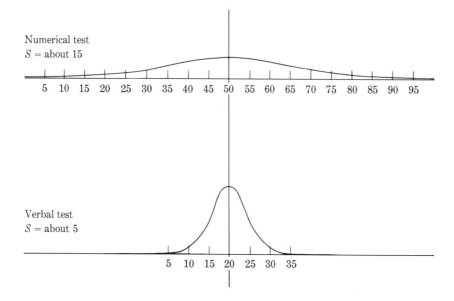

Figure 4-2 Two distributions with the same *N*'s but different variabilities.

score from that point, as in Figure 4-3. The average (mean) of those differences (without regard to sign) can also serve as an index of variability. If, for example, we chose the mean as our reference point, our index would be an average of the individual differences from the mean and could be obtained via the following formula:

$$AD = \frac{\Sigma|x|}{N} \qquad (4\text{-}1)$$

where AD = *average deviation* and $\Sigma|x|$ = sum of *individual deviations* from the mean of the sample.[3]

The "average deviation" is sometimes used as an index of variability, but it is not common enough to be discussed here strictly for its own sake. Rather, I have included it because it is a direct expression of the basic idea that underlies the most used of

---

3. x (the difference between X and $\overline{X}$) is called a "deviation score." The vertical bars in $|x|$ mean "without regard to sign." Σ is defined on page 13.

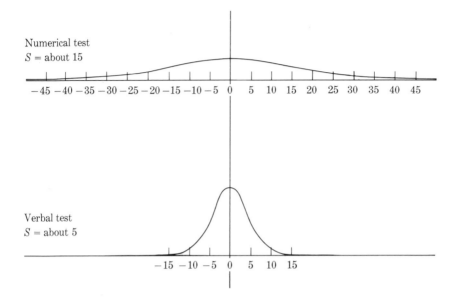

Figure 4-3 The distributions in Figure 4-2 with raw sources converted to a deviation scores.

all measures of variability. That idea is the concept of an *average of individual deviations* from the mean; that statistic is the *standard deviation.*

Now compare the little-used statistic that we have been discussing to the standard deviation (which comes very close to being *the* accepted measure of variability). Here again is the formula for the average deviation:

$$AD = \frac{\Sigma|x|}{N} \qquad \text{(4-1)}$$

and here, the standard deviation: *

$$S = \sqrt{\frac{\Sigma x^2}{N}} \qquad \text{(4-2)}$$

They are very similar, are they not? In fact, they are exactly alike except that for the standard deviation we take a mean of *squared* deviation scores; then we take the *square root* of that mean.[4] But don't worry about the difference between (4–1) and (4–2). The important thing is the similarity: the standard deviation is a kind of *average of individual deviations* from the mean of the distribution.

Appropriate Applications

The standard deviation is to measures of variability what the mean is to measures of central tendency. In a normal distribution, (1) it is the most stable of variability indices, and (2) it lends itself to the computation of other notably stable statistics.

The sampling stability of the standard deviation can be conceived in the same way as that of the mean. (See pages 15–17.) That is, of all variability indices, it varies least among random samples of the same population. The common statistics to which it lends itself are the product-moment coefficient of correlation and the various standard errors, both of which will be discussed later.

The Semi-Interquartile Range

A much easier statistic to comprehend (and to compute) is the *semi-interquartile range*. Look at Figure 4-4. What you see is a negatively skewed distribution cut into four equal parts. At the upper end of each of those quarters is a point on the base line called a *quartile* (Q)—the first quartile (Q_1) above the lowest

---

4. The mean of the squared deviation scores is called the *varance* of the distribution. By now you may have suspected that any set of values of a given kind can be averaged. We have previously dealt with the mean of raw scores (pages 13–14) and of means (pages 15–16). Note that deviation scores, in a normal distribution, meet the requirement of *symmetry* that was set earlier (page 19) for the legitimate use of a mean. The vertical bars in the AD formula mean *without regard to sign*; a deviation score of "– 10," for example, is at the same distance from the mean as a "+ 10." In the formula for the standard deviation, the squaring process has the same effect, because the minus signs drop out. The distribution of such deviations is therefore symmetrical.

Incidentally, in a normal distribution a standard deviation measured from the mean subtends about a third of the area under the curve, and the second standard deviation from the mean subtends about 14 percent. Using that knowledge as a guide, the standard deviations of the numerical and verbal parts of our scholastic aptitude test (Figures 4-2 and 4-3) appear by inspection to be approximately 15 in the numerical distribution and 5 in the verbal. Look at Figure 4-3 and see whether you agree.

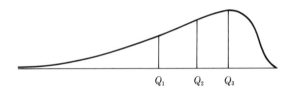

Figure 4-4   Negatively skewed distribution with area divided into quarters.

quarter, the second (Q_2) above the lowest two quarters,[5] and the third (Q_3) above the lowest three quarters. The point just above the highest score in the distribution would logically be Q_4, but that expression is seldom if ever used.

Essence of the Concept
The *semi-interquartile range* is simply the mean of two intervals: Q_1 to Q_2 and Q_2 to Q_3. To compute it, you divide the Q_1-to-Q_3 interval by 2.

Appropriate Applications
The semi-interquartile range is to measures of variability what the median is to measures of central tendency. Although somewhat less stable, it is preferred to the standard deviation in the same kinds of situations in which the median is preferred to the mean—namely, when distributions are radically asymmetric. It is the perfect companion to the median wherever the latter is properly applied.

The Range

There is another measure of variability—one that probably gets more attention than it deserves, both in makeshift analyses and in this book. It is given attention in makeshift analyses because it is so easy to compute, and much of the space assigned to it in this book is occupied by (1) an exposure of its fundamental

---

5. What other familiar statistic has essentially the same definition? If you are not sure, turn back to page 18.

weakness and (2) an attempt to prepare you for possible differences in its interpretation.

Essence of the Concept

Sometimes we are interested specifically in the most extreme cases in a sample; on these occasions we may report its *total range*—or simply *range*—which is the distance from the lowest score to the highest, plus 1. Each of the extreme scores (probably only one at each end of the distribution) occupies a class interval 1 unit long; so, after we have found the distance from the lowest midpoint to the highest, it will be necessary to add .5 at each end in order to include the entire distance from below the lowest occupied interval to above the highest. In Figure 3-1 (on page 15), the lowest score (midpoint of a class interval of 1) is 2; the highest is 9. Subtracting the lowest from the highest would give us 7; but we really should be subtracting 1.5 from 9.5, which would give us a range of 8. Usually the easiest way to compute the range is first to *find the difference between the midpoints* of the two class intervals in which the extreme scores lie and then *add 1.* You may find other textbooks, however, that do not advocate adding the 1. Some authorities regard that refinement as hair splitting because the range is such an unstable statistic in any case.*

In fact, the most important feature of the range is a weakness—its extreme instability. We have just noted that the range in Figure 3-1 is 8. Now turn to Figure 3-4. It is the same as Figure 3-1, with the exception that two scores have been moved away from the main group. Note the effect on the range. (Instead of 8, it is now 35½ – 1½, or 34!) Note, too, that not even *two* extremely high scores were necessary to have that effect; one would have done exactly the same thing. The fact is that the range is determined by *two, and only two, score intervals* in any distribution: the lowest and the highest. That is why it is so easy to compute; that is also why it is so unstable.

Appropriate Applications

We have seen that the standard deviation is analogous to the mean and is its proper companion, and we have noted a similar relationship of the semi-interquartile range to the median. The relationship of the range to the mode is not quite as neat, but there are similarities that should help you to remember both statistics.

One similarity is that each is the *quickest* estimate of its kind. Another, closely related, one is that each is *less stable* than alternative indices. Finally, both the mode and the range are used, more often than are other statistics, to answer questions related directly and very simply to their definitions. For the mode, the question is, "Which is the typical (most frequent) case?" For the range, it is, "How much of the scale must be used to represent the distribution?"

Summary

Two important ways of describing a sample are (1) by its central tendency or average, which indicates the general level of the scores, and (2) by its variability, which tells the extent to which individual scores deviate from that average. This chapter has been about the latter type of description—measures of *variability.*

You have been invited to use your previously developed understanding of measures of central tendency as an anchor for the new concepts; the latter were presented as analogs of the former. Specifically, the *standard deviation* is roughly analogous to the mean and is its companion statistic; the *semi-interquartile range* goes with the median; and the *range* is similar to the mode.

The *standard deviation* is a kind of average of individual deviations from the mean. In a normal distribution, it is the most stable of all measures of variability. It also lends itself to the computation of other stable statistics, like the product-moment correlation coefficient (Chapter 5) and all of the standard errors (Chapter 7).

The *semi-interquartile range* is the mean of the two middle interquartile ranges, or one half of the distance from Q_1 to Q_3. It is less stable than the standard deviation but is preferred to it whenever the distribution is markedly asymmetric, and it is the perfect companion to the median.

The *range* is the distance from the point just below the lowest score to the point just above the highest. Like the mode, it is both easier to compute and less stable than any other measure of its kind. It is, of course, the one statistic that gives information directly about the highest and lowest scores in the sample.

We have seen that in a normal distribution, the mean, the

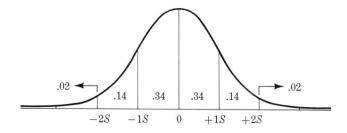

Figure 4-5 Normal distribution with base line divided into standard deviation units.

median, and the mode are all at the same place. When we make a similar graphical comparison of measures of variability, the situation is not quite as simple as that.

In Figure 4-5, the base line is divided into equal units—namely standard deviations—and the areas subtended by them are *un*equal. The number above the line indicates the proportion of the total area that is subtended by each segment. The proportions have been rounded off because they are easier to remember in that form, and they are precise enough for our purposes in any case; more exact proportions are given in Figure 6-2.

In Figure 4-6, the base line is divided into unequal segments, and it is the parts of the area that are equal. And of course the range subtends the entire distribution, as in Figure 4.7.

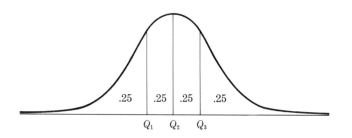

Figure 4-6 Normal distribution with area divided into quarters.

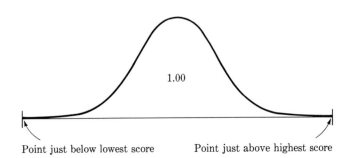

Point just below lowest score   Point just above highest score

Figure 4-7   Normal distribution showing total range on base line.

SAMPLE APPLICATIONS
FOR CHAPTER 4

EDUCATION

You are appointed to a committee of elementary school teachers and administrators. You (the committee) decide to make the teaching of reading a top priority for the next two-year period. You get permission to use the annual inservice training fund for the purchase of a reading program in which all elementary teachers would participate. After an extensive search, you narrow the field to two programs that have been used and evaluated by a large number of elementary teachers on usefulness and practicality. The means of their ratings of the two programs are approximately the same. Is there any other statistic that might help you make your decision?

POLITICAL SCIENCE

You are interested in the incidence of military coups in Latin America. Specifically, you want to know whether most Latin American countries have experienced close to the mean number of coups for the region. After you have gathered your data, how do you obtain that information?

PSYCHOLOGY

You are one of a five-person team of observers sent into a home to evaluate the degree of aggressiveness among family members over a one-week period. The observers indicate that, on the average, eight aggressive acts occur per day. But there is also some disagreement. How might you quantify that disagreement?

SOCIAL WORK

You are the new director of a community fund-raising organization. The member agencies have widely varying needs, but you suspect that recently The Board has been shirking its duty to investigate those needs. Specifically, you suspect that recent allocations have not been sufficiently differentiated. How might you document your case?

SOCIOLOGY

In order to make some decisions about family dwelling construction in your state, a construction firm asks you to report on the variability of family size. You draw a random sample from the state's data on family size. What statistics do you report?

5 | MEASURES OF RELATIONSHIP

There are many times, both in basic science and in professional practice, when we want to know the *relationship* between one thing and another. Indeed, all of science is concerned with such relationships; without knowledge of them, professional practice could never check up on itself.

Take our scholastic aptitude test, for example (pages 12–13). If we merely assume that it is doing what we want it to, then much money and even more time and effort may be spent in vain. If, on the other hand, we define the test's effectiveness in terms of its prediction of success in college, then we have a way of checking up on it. Once we discover how closely the scores are related to some criterion of success, we can judge the test's effectiveness.

What we need, then, is an index of relationship—a number that when low indicates a low degree and when high a high degree of relationship between two variables. We need a *coefficient of correlation.*

Imagine now that the two variables in which we are interested are (1) the height and (2) the weight of a population of toy soldiers. Imagine further that all of those soldiers are exactly the same *shape* and that they differ in size, and therefore in weight.[1] Figure 5-1 pictures a representative sample of them. If we were actually doing the computation, we should need a much larger sample than five, but for learning the concept, the smaller number is better.

---

1. This is, of course, an unnatural situation; I am using it because it enables us to focus our attention exclusively on the two variables of our present concern.

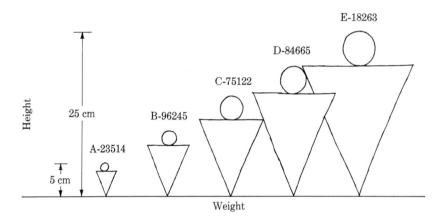

Figure 5-1 Sample of population in which height and weight are correlated positively.

Just by looking at the sample, what would you say the relationship is between height and weight? You notice immediately that the short soldiers are light in weight and that the tall ones are heavy. How would you describe that relationship? Strong? Directly proportional? Perfect? If you were asked to place the strength of the relationship on a scale from .00 to 1.00, you would have to place it at 1.00, because it *is* perfect.*

A coefficient of correlation provides just such a scale—that is, a scale with limits of .00 and 1.00—except that it carries information not only about the *strength* of the relationship but also about its *direction;* some correlations are positive and some are negative.

There are several kinds of coefficients. The easiest to comprehend is the rank-difference coefficient; the most useful (and most used) is the product-moment coefficient. We shall examine each of these in turn.

The Rank-Difference Coefficient (ρ)

Near the beginning of the chapter on measures of variability, I introduced you to a statistic that you will probably never see again. I did so because that statistic, the "average deviation," was the best device available for building an understanding of the concept that lies behind the most-used of all measures of variability, the standard deviation. What I am about to do with correlation is not quite as drastic as that, for the *rank-difference* coefficient is used quite often. But it appears in the literature much less often than the *product-moment coefficient*, and again I am presenting the less frequently used statistic first because it illustrates more directly the essential nature of the more common one.

Positive Correlation

Table 5-1 shows how the data would be ordered in preparation for computing a Spearman *rank-difference* coefficient of correlation on the toy soldiers in Figure 5-1.

In Table 5-1, the first three columns list each soldier's (1) serial number, (2) rank on the variable *height*, and (3) rank on the variable *weight*. The fourth column shows the difference between the two rankings. If you will draw a straight pencil line from X1 to Y1, another from X2 to Y2, etc., the lines will form a ladder pattern. (In a moment, we'll look at a similar table that forms a different pattern.) Notice, too, that *all the rank differences are zero*. Keep that in mind while looking at the formula for the ρ (*rho*) *coefficient:*

$$\rho = 1 - \frac{6\Sigma D^2}{N(N^2 - 1)} \qquad \textbf{(5-1)}$$

where D is the difference between the two ranks of a single subject and N is the number of subjects—that is, the number of pairs of measurements. We are not interested in the computation as such;* the formula is included here only to demonstrate that the rank differences are in the numerator of a fraction that is subtracted from 1.00. That means that the larger the rank differences, the smaller the rho, down to a limit of .00. (After that, still larger differences contribute to rising *negative* coefficients, up to a limit of −1.00.)

Any correlation coefficient carries information about two aspects of a relationship: (1) its *strength*—measured on a scale from zero to unity—and (2) its *direction*—indicated by the presence or absence of a minus sign. Consider again our example in Figure 5-1. I had asked you to imagine that all of the subjects were exactly the same shape, and I attempted to draw them all the same

Table 5-1   Ordering of Data for Rank-Difference Correlation of Heights and Weights of Toy Soldiers in Figure 5-1.

Name or other identification of subject	His rank on variable X (height)	His rank on variable Y (weight)	Difference between the two ranks	Square of the difference
E-18263	1	1	0	0
D-84665	2	2	0	0
C-75122	3	3	0	0
B-96245	4	4	0	0
A-23514	5	5	0	0
Σ				0

shape. But my draftsmanship falls short of perfection (only slightly, mind you!), so if we were to cast those toy soldiers in molds made directly from my drawings, the relationship between height and weight would *not* be perfect, after all. You might expect a coefficient of correlation to be sensitive to that difference between perfection and near perfection; but the rank-difference coefficient is not. Examine Figure 5-1 again, and you will see that all the subjects clearly are ranked the same even when the imperfections of my drawing are recognized. I mention that here because I want to show you later an index (the product-moment coefficient) that *does* take those imperfections into account.

As for the negative sign that accompanies some coefficients, there is an important point to be made (or rather, emphasized, for it has already been made; see above). *The size* (strength) *of a coefficient is entirely independent of its direction* (positive or negative). Although the formula for the rank-difference coefficient does not encourage it, we should think not of a single continuum from negative 1.00 through .00 to positive 1.00 but rather of two separate dimensions—one negative, one positive—each of which begins at .00 and ends at 1.00. A correlation of 1.00 is no bigger (stronger) than a correlation of − 1.00.

Negative Correlation

But what does it mean to say that two variables are *negatively* correlated? Let us look again at the variables of height and weight, but this time in a population quite different from the toy soldiers that we examined in Figure 5-1 and Table 5-1. Consider now a human population in which the shortest individuals are the heaviest and the tallest are the lightest. (Such an arrangement is beyond your experience, but hopefully not your imagination.) Figure 5-2 is an attempted rendering of a sample of five drawn from such a population. The corresponding table (5-2) reveals the "different pattern" I promised a moment ago. Now if you draw pencil lines between equivalent ranks, you will find that a star pattern emerges instead of a ladder.

If you sum the squared rank differences, you will find a marked contrast with the zero that comes from Table 5-1. (Remember that D^2 is in the numerator of a fraction that is *subtracted* from 1.00 to obtain the correlation coefficient.) But my drawing is a considerably less adequate representation of a perfect relationship than it was in Figure 5-1. Does the rho coefficient detect my inadequacy this time? No. Nothing will affect it until the *ranks* change. In Table 5-2, the correlation of height to weight is a full -1.00, even though the weights of Darrell and Erwin, for example, are perceptibly farther apart than their heights.[2]

In other words, information is lost when we convert interval measurements into rank orders. That difficulty can be overcome, at considerable cost in computational labor, by using a different coefficient of correlation. The next section will aim at an understanding, without computational labor, of that coefficient.

The Product-Moment Coefficient (*r*)

The rho coefficient is included in this book primarily because it makes a better vehicle for teaching than the more difficult Pearson *product-moment coefficient.* You will see rho reported from time to time in the literature, but usually it is used only (1) as a relatively quick approximation of the "Pearson *r*," as the product-moment coefficient is often called, or (2) in situations in which the assumptions for *r* cannot be satisfied.*

---

2. That difference may not be as perceptible to you as it is to me, so let me give you an exaggerated example. In the following diagram, the arrangement of Subjects A, B, and C on Variable *X* is quite different from that of the same individuals on Variable *Y*; but their ranks are the same on both.

Variable *X*	A		B		C
Variable *Y*	A	.		B	C

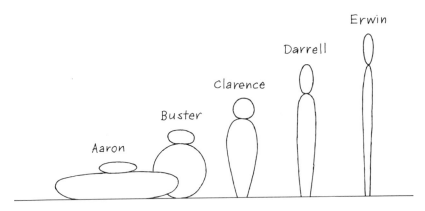

Figure 5-2 Sample of population in which height and weight are correlated negatively.

The Meaning of "Product-Moment"

To begin our discussion of this more sophisticated index, let us look again (Figure 5-1) at those toy soldiers. Once again, we construct a table, and indeed it is of the same general form as Table 5-1; but this time (Table 5-3), instead of *ranks*, we enter ordinary interval scores that tell also how far apart the subjects are on each variable. (See the footnote on the preceding page.)

Don't be intimidated by that table. One reason it is so large is that information must be collected for the computation of two standard deviations. (The x^2 and y^2 columns are used exclusively for that.) The reason for including that computation will come later;[3] at the moment, it is sufficient merely to note the presence of S_x and S_y in the formula for r:

$$r = \frac{\Sigma xy}{NS_xS_y} \tag{5-2}$$

The main idea behind this formula is that when X and Y scores are arranged "in parallel," so to speak, the product of the corresponding deviation scores (xy) is maximally large. That is so

---

3. You will be in a better position to understand that reason after you have been introduced to the concept of "standard scores" in Chapter 6.

Table 5-2 Ordering of Data for Rank-Difference Correlation of Heights and Weights of Men in Figure 5-2.

Name or other identification	His rank on variable X (height)	His rank on variable Y (weight)	Difference between the two ranks	Square of the difference
Erwin	1	5	-4	16
Darrell	2	4	-2	4
Clarence	3	3	0	0
Buster	4	2	2	4
Aaron	5	1	4	16
Σ				40

because in such an arrangement the largest deviation scores—both positive and negative—occur together in the table and thus are multiplied together. Compare the two parts of Table 5-4 and you will see what I mean.

Close examination of the two parts will reveal that the scores are the same in both but that they are arranged differently. On the left, the arrangement is perfectly ordered, as the drawing of toy soldiers on page 38 would suggest; on the right, the arrangement of the Y scores is random, which means that the relationship between X and Y is also random.

Now see what that does to the correlation coefficient. On the left, every large positive deviation score gets multiplied by another that is similarly large and also positive; whereas on the right, a large positive deviation may be neutralized by a low multiplier, and *any* deviation may be multiplied by one of an opposite sign, thereby yielding a negative product. The result is that with the random arrangement, the sum of the *cross products* (Σxy)—and hence the correlation coefficient—is always small. (In fact, it varies from *zero* only by chance!) In terms of the "weightless beam" concept introduced on page 14, you could say that the deviation scores on the two variables are often either close to the fulcrum or on opposite sides of it; whereas in a more substantial correlation, x and y scores for a given individual are consistently together—frequently far from the fulcrum—so that they can apply an extremely large torque, or *moment*, to the beam. Hence

Table 5-3 Ordering of Data for Product-Moment Correlation of Heights and Weights of Toy Soldiers in Figure 5-1.

Serial number	X Height in centimenters	Y Weight in grams	x	y	x²	y²	xy
E-18263	25	250					
D-84665	20	160					
C-75122	15	90					
B-96245	10	40					
A-23514	5	10					

the term "product-moment"—a reference to "the product of the moments," which is largest when all of the scores are perfectly ordered.

The Scatterplot

There is another way of looking at correlation that gets directly at the fundamenal idea. I refer to the *scatterplot*. A scatterplot is a surface on which both the X and the Y scores of each individual can be represented by a single point. If, for example, we were correlating scholastic aptitude test scores with first-year grade-point averages and if every student's test score were precisely proportional to his subsequent scholastic performance, the scatterplot might look like Figure 5-3. Each student is represented by a single point (represented here by a black dot) that locates him on two dimensions, X and Y. In this case, all of those points lie in a straight line (called a *regression* line because it embodies the mathematical regression of Y on X.)[4] The "scatter" that we have

---

4. The slope of a line in a coordinate plot like this is the ratio of the amount of change on the ordinate (Y) to the amount of change on the abscissa (X). If Y increases ½ unit for every unit increase in X, the slope is .5. If while following a regression line with your pencil you find that you have to move four units on the X axis for every one you cover on Y, the slope is .25—providing that the changes on both variables are positive. The same *degree* of slope could occur in a negative *direction* if Y were *dropping* ¼ unit with every one-unit rise in X. The relation between scatter and slope is explained later (page 58), after you have mastered the concept of the standard score. For an interesting historical account of the development of regression and correlation concepts, see J. P. Guilford and B. Fruchter, *Fundamental Statistics in Psychology and Education*, 6th ed. (New York: McGraw-Hill, 1977).

Table 5-4   Effect of Ordering on Sum of Cross Products (Σxy).

	Ordered						Random			
X	Y	x	y	xy		X	Y	x	y	xy
25	250	10	140	1400		25	160	10	50	500
20	160	5	50	250		20	40	5	−70	−350
15	90	0	−20	0		15	10	0	−100	0
10	40	−5	−70	350		10	250	−5	140	−700
5	10	−10	−100	1000		5	90	−10	−20	200
$\frac{\Sigma}{X}$ 75	550			3000		75	550			−350
15	110					15	110			

been discussing determines the slope of the regression line; the
smaller the scatter, the steeper the slope and the bigger the correlation. In our illustration, there is *no* scatter, which is how it
always is with a perfect correlation.

Notice, however, that at least some of the students in our
example *could* deviate perceptibly from the regression line without changing any *ranks* on either variable. Select some adjacent
individuals (at the upper or lower ends of the line, because they
are farther apart there) and try it a few times with your pencil.
You will find that within the "same rank" limitation, there is
considerable freedom to change measured values on *X* and *Y*. The
amount of such freedom is the amount of information that is lost
when interval scores are converted to ranks. That information is
retained when the correlational statistic is a Pearson *r*.

If, as in real life, the correlation is not perfect, there is always
some scatter. But if the *X-Y* relationship is extremely strong, the
deviations are small (see Figure 5-4); we can make very accurate
predictions of *Y* scores from a knowledge of *X* scores. At the other
extreme is the "random" *X-Y* relationship depicted in Figure 5-5.
There, there is no tendency for a low *Y* to be associated with a low
X or a high *Y* with a high *X*, and as you can see, there is a near
maximum of scatter. Knowing a person's score on *X* does not help
at all in estimating his score on *Y*.

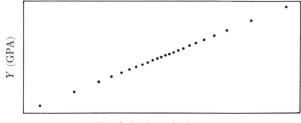

Figure 5-3   Scatterplot of perfect positive correlation.

Negative correlations behave in exactly the same way, except that the regression line slopes *down* from left to right instead of up. Figures 5-6 and 5-7 are examples.

Figures 5-3 through 5-7 have been contrived to represent perfect positive, high positive, zero, high negative, and perfect negative correlations. Either of the "perfect" correlations would enable us to predict Y precisely from a knowledge of X, and vice versa. The zero correlation would tell us that a knowledge of an individual's position on one variable would be useless in predicting where

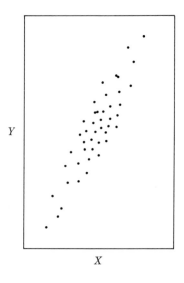

Figure 5-4   Scatterplot of high positive correlation.

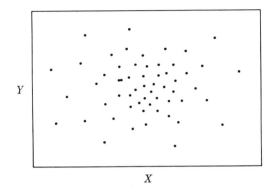

Figure 5-5   Scatterplot of zero correlation.

he is on the other. Such a correlation would say that high scorers on the scholastic aptitude test are just as likely to flunk out of school as low scorers. The other correlations (high positive and high negative) would not insure that we could make perfect predictions, but they would significantly reduce our errors, should we wish to try predicting college performance from test scores.

Figure 5-8 depicts those relationships in a slightly different way. Of the four lines that cross diagrams A, B, and C, one line (the perpendicular) is the mean of all the X scores, another line (the horizontal) is the mean of the Y's, and the third and fourth lines (the diagonals) are regression lines. The reason that you don't see

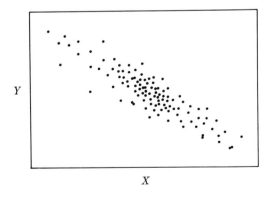

Figure 5-6   Scatterplot of high negative correlation.

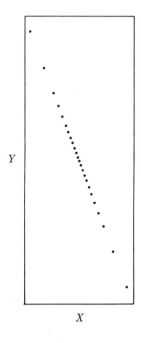

Figure 5-7   Scatterplot of perfect negative correlation.

four lines at $r = 1.00$ (diagram A) and $r = .00$ (diagram C) is that when one line corresponds exactly to another, you can see only one. In diagram A, where the correlation is perfect, the regressions of y on x and of x on y are merged into one line that is just 45° from each of the two base lines. In diagram C, where $r = .00$, the regression lines are separate, but the regression of Y on X is at the mean of Y all the way across, and that of X on Y is at the mean of X; so the two regression lines are superimposed on the two mean lines. (They *are* the two mean lines.) Thus, you see two lines instead of four.

Have you noticed that in contrast to Figures 5-3 through 5-7, all of the scatterplot grids in Figure 5-8 are precisely square? There is a very good reason for that, as you will see later; but it cannot be explained until you have mastered the concept of the "standard score." Watch for that concept when you read Chapter 6, and after you've mastered it return to get another perspective on Figures 5-3 through 5-8.

There is one feature of Figure 5-8 that could be misleading if not accompanied by some explanation. It is especially noticeable

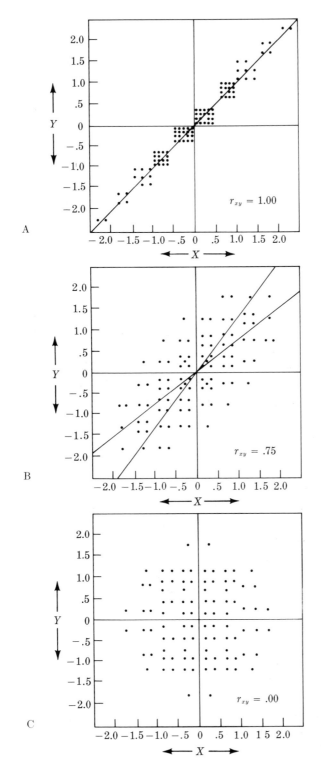

Figure 5-8 Three scatterplots on standard grids (all scores in standard units).

in diagram A in which it appears that there is, after all, some scatter when the correlation is perfect ($r = +1.00$).

That is an erroneous impression. What has happened is that dots that should have been stacked directly on top of each other were dispersed horizontally so you could see them. Figure 5-9 is an attempt to represent three dimensions. You can see there that when the correlation is perfect, there is *no* scatter. In stark contrast, a zero correlation spreads out over most of the board. Other correlations approximate those two in varying degrees.

Effect of Restricted Variability

Everything I have said about correlation is based upon the assumption that the individuals measured constitute random samples from both distributions. If the correlation is between IQ and scholastic achievement, for example, we should select a large sample of subjects and measure each of them twice—once for IQ and once for achievement. Their scores on the IQ test should form a distribution similar in every way but size (N) to the one that would have emerged had we tested an entire population in the same way; and similarly for achievement.

On the other hand, we saw very early (pages 12–13) that it is sometimes useful to subdivide a general population into subcategories and that it is legitimate to call each subcategory a "population," too. For example, instead of including literally everybody in a population of IQs, we can arbitrarily define a population that includes only college students. Now our "sample" must be taken at random * from *that* "population."

All of this is fine as far as it goes, but we must recognize that it has a very significant effect on the correlation coefficient. The best way to demonstrate that fact is again to draw a diagram; Figure 5-10 is appropriate to our present need. It shows a hypothetical relationship between intelligence (as measured by a standardized test) and achievement (as measured by grade-point average) in the general population. It also shows what the relationship would be if the population were defined as "applicants to graduate schools." The entire range of IQ scores for the general population forms the base line of the diagram, and the entire range of grade-point averages forms the ordinate. Brackets I and II enclose those segments on each scale within which 90 percent of applicants to graduate school may be found.

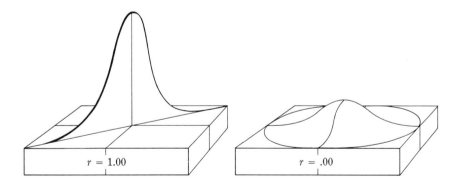

Figure 5-9 Three-dimensional rendering of two scatterplots.

The scatterplot in Figure 5-10 reveals a substantial correlation between intelligence and scholastic achievement. However—and this is the point toward which this discussion has been directed—a plot of the area marked off by segments I and II reveals no relationship at all. Figure 5-11 is an enlargement of that area.

Be cautious, then, whenever you interpret a low correlation coefficient. In the above illustration, it would be proper to say that among applicants to graduate schools, there is virtually no relationship between IQ and GPA; but be careful *not* to infer from that

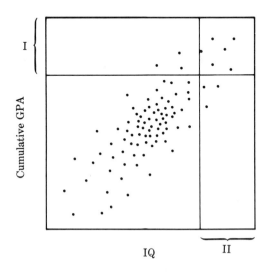

Figure 5-10 Scatterplot of high positive correlation, showing effect of restricted variability.

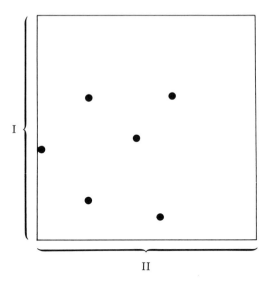

Figure 5-11   Restricted area of scatterplot (enlarged).

a similar lack of relationship in the general population. A correlation coefficient is relevant only to the population from which the sample has been drawn.

Correlation and Causality

Before we can be certain that we have found an instance of *causation*, we must observe a reliable relationship between the purported causative agent and the event being caused. If every time you push a switch in your bedroom to the "up" position a light goes on, you come to believe that the latter event is caused by the former, even in the absence of wiring diagrams, electrical theory, and so forth. If your dog becomes wildly aggressive whenever the garbage man approaches, you say that the man's approach is causing the dog to be upset. If in studying a large sample of school children you find that those with larger vocabularies have higher IQs, you assume that the vocabulary is a cause of the intelligence.

But should you? Is it legitimate for you to assume that when two events occur together one is caused by the other? Consider the following case: what size of coefficient would emerge if we were to compute a correlation between the revolutions per minute of the left and right front wheels of your car (Figure 5-12)? The

Left

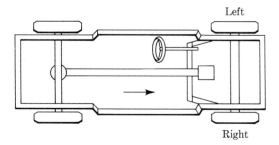

Right

Figure 5-12   Schematic automobile chassis.

correlation would be pretty high, would it not? If the car were on a curving road and if turns in one direction were predominant throughout any given minute, the two scores would not be the same for that minute; but with varying speeds on a straight road, the correlation would be almost perfect (Table 5-5).

But does the speed of one wheel *cause* that of the other? Only to the extent that each is a functioning part of a total system in which the two events occur. It would be more accurate to say that the functioning system—the car in motion—causes the revolutions of *both* wheels.

I don't suppose anyone will ever bother to find the correlation between the two front wheels of his automobile, but the same principles apply also to many situations in which correlation coefficients *are* commonly computed. For example, if we were to study a large sample of elementary school children, we might find a tendency for those children who display extensive vocabularies to be superior students in arithmetic. We might be tempted to conclude that a good vocabulary causes competence in arithmetic. But then we find that chronological age correlates with both variables. Might it not be better to say that attainments in both vocabulary and arithmetic are "caused by" (are parts of the functioning system of) the child's developing intelligence?

The caution that I am advocating here has important practical implications; for without it we might easily be persuaded, for example, to set up elaborate vocabulary-training programs for children who show weakness in quantitative thinking. Some wag once noted that there is a substantial correlation between the intelligence of boys (as indicated by their mental ages) and the length of their trousers. He suggested that a relatively inexpensive

Table 5-5   Arrangement of Data in Correlation of RPMs over Minutes of Travel.

Successive minutes of travel	Revolutions of left wheel	Revolutions of right wheel
1	?	?
2	?	?
3	?	.
.	.	.
.	.	.
.	.	.

way to increase the intelligence of boys would be to increase the length of their trousers!

So although correlation is a *necessary* feature of a causal relation, it is not *sufficient* to prove that a causal relation exists. Whether the relation is interpreted as a causal one should depend not just on the correlation of two variables but also on some rational link between them—on the extent to which the relationship "makes sense" within some sort of conceptual framework (within a wiring diagram, for example, or a sociological theory).

Correlation is not causation.

Reliability and Validity

At the beginning of this chapter, I suggested that an index of correlation would allow us to check on the accuracy of predictions made by a scholastic aptitude test. Now, by substituting the test score for X and college GPA for Y in each of the diagrams in Figure 5-8, you can see just how accurate the predictions would be if the correlation were .00, if it were .75, and if it were 1.00.

A correlation coefficient that is used, as in the case of the above-mentioned scholastic aptitude test, to predict a *criterion* score (Y) from a test score (X) is called a *validity coefficient.*[5] (The

---

5. The examples given are of "predictive validity" and "test-retest" reliability. There are other kinds of validity and of reliability, but these examples serve to illustrate how a correlation coefficient can serve as an index of either.

criterion is the entity that our test is supposed to be measuring—in this case, "scholastic aptitude," or the capacity to profit from schooling.[6] If we were to administer our scholastic aptitude test to all entering freshmen as indicated earlier, and if we were to test them again a week later, we could compute an r that would tell us the relationship between the first and second testings. Instead of an r_{xy}, we would have an r_{xx}, and it would be called a *reliability* coefficient, because it would tell us the extent to which we can depend on the test to give us the same results from one administration to the next. (You can't measure anything reliably with a rubber yardstick!)

The "rubber yardstick" metaphor is a good one to help you remember the concept of reliability, but it bears no direct relation to validity. Here is an analogy that illustrates both reliability and validity:

Imagine a man aiming a rifle in the general direction of a target that is several hundred feet away from him. I say "in the general direction" because there is a huge sheet of white gauze that hides not only the target but a considerable area around it. Our marksman takes aim, then, at a spot that he hopes is the bull's eye. He fires five shots—and scatters them all over that cloth screen. Figure 5-13 shows the pattern that they make on the screen and on the hidden target.

Another marksman takes aim at the same spot as the first one did. He "groups" his shots as shown in Figure 5-14, thereby missing the target completely.

And finally, a third marksman decides where *he* thinks the target is, aims at its supposed center and pumps off five shots that are as close together as those of the second marksman and are also perfectly placed on the target as in Figure 5-15.

Now review the concepts of reliability and validity, and see if you can tell which of them is illustrated by which marksman. Reliability? That would have to be the closely grouped patterns, wouldn't it? Shot after shot lands in nearly the same place; you can count on it, just as you can count on a reliable test to yield

---

6. GPA is generally accepted (though reluctantly by some) as an index of scholastic achievement—i.e., the extent to which an individual has "profited from schooling."

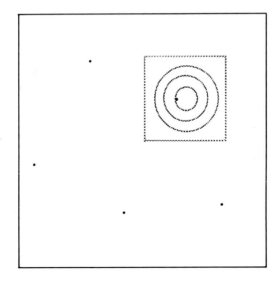

Figure 5-13 Sheet of white gauze with invisible target; shots widely spaced.

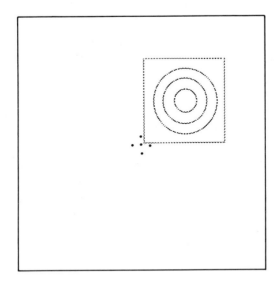

Figure 5-14 Sheet of white gauze with invisible target; shots grouped closely together off target.

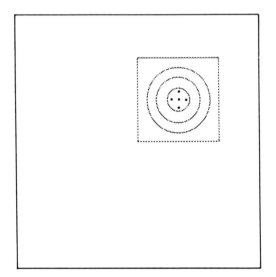

Figure 5-15 Sheet of white gauze with invisible target; shots grouped closely together on target.

nearly the same scores when the same people are retested. Proximity to the target, on the other hand, is the analog of validity. Close grouping of shots doesn't help a bit unless it is on target. Reliable testing is useless unless we know what it is we are testing.

So high reliability does not insure high validity. But it is also important to note that *low* reliability absolutely precludes it. Unless your shots are all very close together, it is impossible to make a perfect score. If a test has low reliability, it can have *some* validity (see Figure 5-13) but not much.

To put it another way, a test can be very reliable without being valid, but it can't be very valid without being reliable.

If when you have finished this book you continue your study of measurement, you will have an opportunity to develop a better understanding of the kinds of validity and reliability that I have barely touched upon; you will also discover that there are others that I have not even mentioned. We will not go further into those matters here, but do be alert to later applications. Even before you finish the book there will be opportunities to learn more about validity and much more about reliability. The word "reliability" won't be used very much, but Chapter 7 is mostly about "precision" of measurement, which is really another way of saying "reliability"; and Chapters 8 and 9 are concerned with the "significance" of obtained differences, which also answers the question, "How reliable is the information that we have obtained?"

Summary

It is often important to know the relationship between two variables. Coefficients of correlation are indices of relationship. Many such indices have been devised for various applications, but we have discussed just two: the rank-difference and the product-moment coefficients.

Any *correlation coefficient* is an index of the extent to which measurements of the same individuals are to be found on corresponding segments (low, middle, high) of two different scales. When such a relationship holds precisely and with no exceptions, the correlation is said to be perfect, and the coefficient is 1.00. Less than perfect relationships produce coefficients smaller than 1.00. A *negative* correlation is just as strong as a *positive* correlation if its coefficient is equally large; the coefficient indexes magnitude and direction separately in a single number that (1) varies from .00 to 1.00 and (2) does or does not carry a minus sign.

The Spearman *rank-difference* (*rho*) procedure places all subjects in order from smallest to largest on the basis of their scores on X; then it ascertains how nearly their Y scores approximate that order. The better the approximation, the smaller are the differences in rank (hence the name "rank difference") and the larger the coefficient of correlation.

The Pearson *product-moment coefficient* (*r*) is like rho except that it retains information that rho throws away—namely, the distances that separate subjects of adjacent ranks on either of the variables being correlated.

If the position of every subject is plotted in terms of both X and Y variables, a pattern called a *scatterplot* emerges. In that pattern, it is possibly to perceive directly the degree of relationship between the two variables.[7] The same diagram reveals the direction of that relationship.

---

7. We might still want to compute *r*, however; the most *direct* index of an entity is often not the most *precise* one.

Correlation procedures have many uses in the technology of testing, but two are particularly noteworthy. In the kinds of situations cited here, the *coefficient of reliability* tells us whether a test does *consistently* whatever it does; the *coefficient of validity* tells us whether it is doing what we have *believed* it to be doing.

The formula for r deals primarily with the products of individuals' deviation scores, but an important part of it is also the two standard deviations in the denominator. An adequate explanation of their presence there awaits your understanding of the "standard score"—a concept that is the core of our next chapter.

SAMPLE APPLICATIONS
FOR CHAPTER 5

EDUCATION

You are an elementary school consultant helping teachers plan activities to enhance students' emotional and social development. You have observed informally that students who have a negative view of themselves (low self-concept) tend not to become involved in helping class groups achieve common goals (low social responsibility). You think that if there is a strong relationship between self-concept and social responsibility it might be useful to plan activities which would enhance development in both areas. To investigate this relationship you administer a self-concept and a social-responsibility scale to a random sample of 200 fourth-, fifth-, and sixth-grade students. After the data have been collected, how do you organize them to answer your question about relationship?

POLITICAL SCIENCE

Researchers have frequently asked whether there is any relationship between (X) the amount of domestic conflict within a given country and (Y) the amount of foreign conflict which that country initiates. Assume that you have constructed a conflict scale and collected data for fifty countries on the values of both X and Y. What statistic will answer your question?

PSYCHOLOGY

You are a child psychologist interested in the possible positive relationship between the amount of sugar in the breakfast diet of young children and hyperactivity (i.e., excessive muscular activity, inattentiveness, etc.). You ask 100 mothers of elementary school children to keep records on what their children eat and drink each morning for breakfast. A nutritionist then analyzes the parental reports and notes the average weekly amount of sugar ingested. Data on hyperactivity are gathered by behavior analysts who visit classrooms on a daily basis and rate each child on a ten-point continuum. What statistic will indicate the strength and direction of the relationship between sugar intake and activity level?

SOCIAL WORK

You are a social worker studying the records of women who receive AFDC payments. You are interested in the possible association between a woman's level of self-esteem and the number of months she receives benefits. What single index can summarize information pertinent to this problem?

SOCIOLOGY

Your research class is locked in a disagreement over the possible relationship between the conservatism or liberalism of academic discipline and authoritarianism. The question is whether liberal arts disciplines like English, history, philosophy, and psychology either attract or produce professors who are significantly more liberal and less authoritarian than professors of disciplines like math, biology, chemistry, business, and physics, who, some students claim, are more conservative and authoritarian.

Choosing this issue as your research topic, you collect data from professors on your campus using the Authoritarianism Scale to determine authoritarianism. To identify the conservative-liberal orientation of academic disciplines, you pick a panel of ten judges to place the disciplines (not the professors) on an interval scale from 0 (most conservative) to 100 (most liberal).

How do you quantify the relation between these two variables?

6 | INTERPRETING INDIVIDUAL MEASURES

Measurement has been defined as "rules for assigning numbers to objects to represent quantities of attributes."[1] The importance of *rules* lies primarily in the fact that the resulting "assignment of numbers" nearly always functions as a *communication*, and unless the recipient of a communication knows the rules by which the sender has made that assignment, he will be unsure of its meaning. Many such rules are intuitively obvious. If you tell me that you have measured the distance from Point *A* to Point *B*, I shall assume that the rule you followed was to lay a ruler across the two points and note how many units (for example, inches) lie between them.

But most rules are not so obvious. What rule do you follow if you want to tell me the size of a particular circle? Do you lay your ruler across the center and the circumference and report the number of units between the two? If so, you had better label the resulting number as a "radius," because then I shall know exactly what you did to get it. There are, of course, two other labels— "diameter" and "circumference"—that you could use and that would specify two rather different operations.

Those are actually alternative operations for measuring a circle, but they have become standardized over the years so that their descriptions are no longer necessary to precise communication; mere labels suffice. In some educational and psychological mea-

---

1. Jum C. Nunnally, *Psychometric Theory* (New York: McGraw-Hill, 1967), p. 2. [Copyright 1967 McGraw-Hill. Used with permission of McGraw-Hill Book Company.]

surements, a similar standardization has occurred—though it is not as complete because the operations are much more complex. "Level of intelligence," for example, can be represented by a number, but the number is rendered more meaningful by a specification of the particular standardized test from which it came; and even then, the specificity of the operations falls far short of that of the simple physical examples described in the previous paragraph.

Other measurements are not standardized at all. If we want to do a rat study in which one of the variables is "drive level," we have to figure out a way to measure that level; then, having done so and run our experiment, we are obliged to describe in some detail the operations that resulted in whatever scores (measures) we have to report. It is in situations like this one that the need for specifying rules is most apparent; but the rules are there even when, as in the measurement of a circle, they are not explicitly stated. Were it not so, there could be no communication.

Our definition says that the rules we have been discussing are "for assigning numbers to objects. . . ." It might be worth a couple of sentences to point out that numbers are not always so used. A mathematical system (a set of rules for the manipulation of symbols) can make use of numbers without any reference to objects. Such a system is viable to the extent that (1) its operations are internally consistent and (2) there is no requirement that it bear any relation to the real world.

So numbers are not always assigned to objects. In measurement, they are so assigned. But even in measurement, a number does not represent an object as such, but rather, the quantity of some *attribute* of the object. How smart is this boy? How tall is that toy soldier? How long is the segment *A–B* on the surface of this burial site? The numbers that we assign to the objects represent the magnitudes of their attributes.

The definition of measurement quoted on the previous page is adequate as far as it goes; it is a good general definition. But for our particular purpose, one further point needs to be made: in many measurement problems, scores can legitimately be interpreted only as *individual differences*. Even in a notably sequential discipline like mathematics, when your teacher tells you that he is grading his students "on absolute standards," there is reason to suspect that his statement is not entirely correct. He may require that first-year work be mastered before he "passes" you to the second year; but he must have some idea, either from his own experience or that of his mentors and colleagues, what can reasonably be asked of first-year students. His "absolute standards" turn out to be less absolute than he thought.

The example is from education. I am not saying that it is never possible to make any educational measurement without taking into account many other measurements on the same dimension. What I am saying is that in most applications, the more those other measurements are utilized in the interpretation of an individual score, the more sophisticated the interpretation will be.

Consider the case of Joseph O. Cawledge. JOC's high school grades were pretty bad; but then he was a full-time athlete, and the rest of the school's program never interested him very much. How well might he perform scholastically if he were motivated to do so? To find out, we give him Cerebral University's scholastic aptitude test.

You will recall that the CU test is divided into two parts—a numerical test and a verbal test. Joe Cawledge takes them both because they are both required of all entering freshmen. He gets a score of 60 on the numerical test and a score of 30 on the verbal. Now, what do we know about Joe?

It *looks* as though he is twice as good with numbers as he is with words. But mightn't that appearance be deceiving? Turn to page 27 and place each of Joe's two scores within the appropriate distribution in Figure 4-2. (Insert a bookmark there, for we shall be using Figure 4-2 repeatedly for the next few moments. In these as in any interval measurements, two questions must be answered about each scale: (1) what is its *point of reference* and (2) what is its *unit of measurement?* Answers to those two questions about a scale would make it possible for us to answer two parallel questions about any individual score: (1) is it *above or below* the reference point and (2) *how far* is it from that point?

It is customary to use the mean of some well-defined "standardization" sample as the standard reference point. You can see immediately that each of Joe's scores is above that point on the appropriate scale. But how far above? "Ten points," you say? "They are the same distance above their respective means." Look again.

One score (N) is in the populous middle part of its distribution, and the other (V) is way out in the "tail" of its distribution. On second thought, then, doesn't it seem to you that the latter must be a much higher score than the former?

Yes, but how much higher? Once again we find that drawings are excellent aids to a comprehension of basic principles, but for precision we need a numerical index. We need a single number that will tell us how far away from the mean any given score is and in what direction.

Standard Scores: The *z* Scale

The answer to the question "how far?" must be in terms that make it possible to *compare* an individual's score on two different scales. To put it another way, we must find a way to put both scores onto a single scale—a *standard* scale, if you will. That probably doesn't mean much to you right now, but in a moment it will.

We have already found a common reference point for both distributions, have we not? Every distribution has a mean; therefore the mean can be used as the reference point for the standard scale that we are about to develop. Every score that falls precisely on the mean of its own distribution will be given a score of zero, because all counting of units will start from there.

Now that we have a common reference point, all that remains is to find a common *unit* for our new scale. As Figure 4-2 demonstrates, it must be a unit that can be used to *compensate for the variability* of a distribution; and in order to do that, it must be a measure of variability. If we were to divide Joe's numerical deviation score by a large number and his verbal score by a small one (look back at Figure 4-2), wouldn't we have a more realistic index of his position in each distribution?

A measure of variability *would* be large for the numerical and small for the verbal distributions, so that is how we shall accomplish our purpose of converting both scores to a common scale. If you had a measurement stated originally in inches and you wanted to convert it to feet, how would you proceed? You would divide by the number of inches in a foot, would you not? Well, you do the same thing here: you divide by the number of raw-score units in a *standard deviation* (the standard deviation of that particular distribution). The result gives you the number of standard deviations in the interval being measured. The formula looks like this:

$$z = \frac{x}{S} \tag{6-1}$$

where z is a standard score, x is a deviation score, and S is the standard deviation of the distribution in which x occurs.

Let's see how it works in Joe's case. If the standard deviation of the numerical distribution is 15, and Joe's score of 60 is ten raw-score points above the mean, his deviation score is 10 and his standard score is 10/15, or *0.67*. If the standard deviation of the verbal distribution is 5, and Joe's score of 30 is again ten points above the mean, his deviation score is 10, but his standard score is 10/5, or *2.00*. Quite a difference, is it not? Figure 6-1 shows where Joe's two scores would be in a distribution of standard scores from both tests.

We have accomplished our purpose: we have moved scores from two different scales onto a single "standard" scale. When each was expressed in terms of its own scale, the two could not be compared; now they can be.

Other Standard Scores

Effective though it is, our standard scale still has some shortcomings in actual practice. For one thing, the negative numbers are a nuisance; for another, the unit is awkwardly large.

Fortunately, however, neither of those shortcomings is difficult to overcome. To get rid of the negative numbers, we have but to add a constant to every z score; if that constant were 5, the mean score would be $0 + 5 = 5$, $1S$ below the mean would be 4, two above it would be 7, and so on. (See line B, Figure 6-2.) To reduce the size of the scale unit requires nothing more than dividing it into smaller parts, which of course makes the *number* of parts larger. If we want our new units to be one tenth the size of the original, there will be 10 of the new units to every standard deviation. Combining those maneuvers (first adding 5 to every z score and then multiplying the result by 10) produces the scale that you see in line C of Figure 6-2 and opposite the label "T scores" in Figure 6-3.*

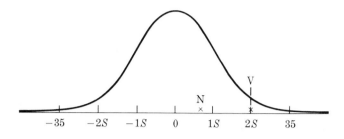

Figure 6-1   Distribution of standard scores from two different tests.

Other scales are constructed in much the same way. In the "College Boards" scale (CEEB in Figure 6-3), 5 again is added, but this time 100 is the multiplier; the resulting distribution has a mean of 500 and a standard deviation of 100. The Army General Classification Test mean is arbitrarily set at 100 and its standard deviation at 20. In every case, what we have is but a modification of what is still basically a standard-deviation scale.[2]

Standard Scores in Correlation

In Chapter 5, footnote 3, I promised that once you have been introduced to the concept of the standard score, you would be ready to develop a deeper understanding of the Pearson product-moment coefficient of correlation. You have now been properly introduced, so let's take another look at correlation.

The reason for the standard deviations that you see in Formula 5-2 is that they are necessary to the conversion of raw scores into standard scores. We may emphasize that conversion by rewriting the formula

$$r = \frac{\Sigma xy}{NS_xS_y} \tag{5-2}$$

Rewritten, it looks like this:

$$r = \frac{\Sigma z_x z_y}{N} \tag{6-2}$$

2. You can afford to ignore the "Stanines" and "Wechsler Scale Subtests," but it would be well worth your while to study the rest of Figure 6-3 until you understand thoroughly all of the conversions that are illustrated there.

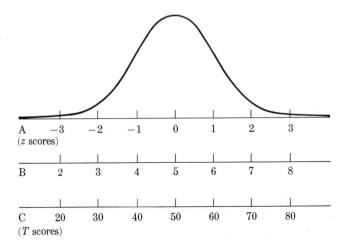

Figure 6-2 Evolution of a standard scale with a mean of 50 and a standard deviation of 10.

If you remember that z scores are in standard-deviation units and also that Σ/N is the way you compute any mean (pages 13–14), it should be clear to you that r is the mean of the cross products of x and y when *both are expressed in standard units*. It is also *the slope of the regression of Y and X*, again when both are in standard units.

If we make scatterplots of ten raw-score relationships that are all of the same strength and direction, we may get ten different slopes of our regression lines; if we first convert all measurements to standard scores, we get identical slopes. Figures 5-3 and 5-7 are positive and negative, respectively, but they are both perfect relationships; if they were plotted in standard units, the degree (though of course not the direction) of their regression slopes would be the same. Without the conversion, they are very different. Figure 5–8A also shows two perfect relationships (Y on X and X on Y), but this time with identical slopes for both; you see only one line. That's because only standard scores are used there.[3]

---

3. If you will turn Figure 5-8 over on its side and examine the regression of X on Y, you will discover that in every diagram it is exactly the same slope as that of Y on X. (The direction of the slope is reversed because the sequence of Y scores is reversed when the diagram is viewed from the side.)

If your plot is done with raw scores, the length of each unit on either dimension is arbitrary. For example, you may use inches or feet on one dimension, ounces or pounds on the other, and you may let one inch, foot, ounce or pound be represented by whatever distance is convenient, depending on the kind of graphing paper you are using and the horizontal and vertical space that is available. Once the scores have been converted, and that is no longer true; for a given strength of relationship, there is one and only one degree of slope. When no relationship exists, the slope is zero; since there is no reason to expect the z_y scores accompanying one z_x value to be any different from those associated with any other, the best estimated central tendency of the z_y's at any given value of z_x is simply the mean of *all* the z_y's in the entire distribution. So as our regression line moves from left to right on the X axis, it moves not at all on the Y; which is to say, the slope of the regression line is zero, as in Figure 5-8C. For a perfect relationship, it is 1.00, as in Figure 5-8A. All other relationships are somewhere between those two extremes.

In short, one meaningful interpretation of r is as *the slope of the regression of Y on X when both are laid out in standard units.*

Centile (or "Percentile") Scores

Standard, or scaled, scores can be used to compare an individual's scores on two different tests or to trace with one test his progress over a period of time. Whatever else it may do, every such score compares the individual's performance to that of a *standardization or "norm" group*. But the size of that group is commonly monstrous and its composition heterogenous. Often what is most needed is a comparison with a group that is less cumbersome and more precisely defined.

Take Joe Cawledge's numerical performance, for example (pages 64–66). Imagine now that the test, far from being homemade by Cerebral University as I have asked you to think of it until now, is only being borrowed by the University as a part of a national testing program. The test has already been *standardized*—that is, it has been administered to a very large number of college students, measures of central tendency and variability have been computed, scaled scores (say T scores) have been derived, and *norm tables* have been prepared that make it possible

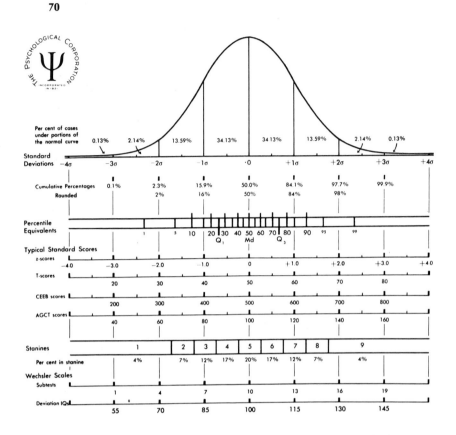

Figure 6-3 "The normal curve, percentiles, and standard scores. Distributions on many standardized educational and psychological tests approximate the form of the *normal curve* shown at the top of this chart. Below it are some of the systems that have been developed to facilitate the interpretation of scores by converting them into numbers which indicate the examinee's relative status in a group.

"The zero (0) at the center of the baseline shows the location of the mean (average) raw score on a test, and the symbol σ (sigma) marks off the scale of raw scores in *standard deviation* units.

"Cumulative percentages are the basis of the *percentile equivalent* scale.

"Several systems are based on the standard deviation unit. Among these *standard score* scales, the z score, the T score, and the stanine are general systems which have been applied to a variety of tests. The others are special variants used in connection with tests of the *College Entrance Examination Board*, the World War II *Army General Classification Test*, and the *Wechsler Intelligence Scales.*

"Tables of norms, whether in percentile or standard score form, have meaning only with reference to a specified test applied to a specified population. The chart does not permit one to conclude, for instance, that a percentile rank of 84 on one test necessary is equivalent to a z score of to convert quickly any individual raw score to other kinds of scores. (You will recall that a raw score by itself is meaningless.) Imagine further that Joe's score on the numerical test is one standard deviation above the national mean.

That is important information in itself, and it makes possible a comparison of Joe's present performance with his performance on the same test months or years later. What it does *not* do is to compare him with the competition at CU; more specifically, it does not compare his performance with the performances of other entering freshmen at Cerebral University. But the national testing program includes the gathering each year of data concerning those performances—data that show up in next year's norm tables in the form of *centiles*—or, as they are more often called, *percentiles*. Actually, of course, the data themselves are raw scores; some kind of transformation must be accomplished in order for the information to be used in this new way.

What is this new way? It is in fact very simple. Do you remember the concept of the "quartile" (page 29)? It is the point on the base line that lies just above a specified *quarter* of a distribution. If the point above a quarter is a quartile, how might we refer to the point above a *tenth* of the scores in a sample? *Decile* would be the parallel term. And what if we were to divide the sample into 100 equal parts? The parallel term there would be *centile*, would it not? A centile is the point on the base line that lies above a specified number of *hundredths* of the distribution. The term *percentile* is technical slang for "centile."

Now back to Joe. His numerical raw score was 60, which happens by coincidence to be precisely one standard deviation above the national mean and converts to a T score that is also 60. That conversion has been done by a computer at national headquarters, and neither Joe nor his counselor ever sees the raw score.

$+ 1.0$ on another; this is true only when each test yields essentially a normal distribution of scores and when both scales are based on identical or very similar groups of people.

"The scales on this chart are discussed in greater detail in *Test Service Bulletin No. 48*, which also includes the chart itself. . . . Copies of this bulletin are available from The Psychological Corporation, 304 East 45th Street, New York, N.Y. 10017" [Psychological Corporation, "Methods of Expressing Test Scores," *Test Service Bulletin No. 48*, September 1954.]

What they do see is a standard score of 60 and a *norms table* that allows them to convert it into a "percentile" in any of several populations. (Samples from those populations are called *norm groups,* and the process of gathering data from them is often referred to as the *standardization* of the test.) Table 6-1 shows how the norms might look for the numerical test. The table is represented graphically in Figure 6-4.

The important thing to notice is that Joe has not one score, but many. If it is true that his performance is meaningless until it has been compared to other performances, it is also true that its meaning is enhanced by a precise definition of the group to which he is being compared at any given time. Once that definition has been made, individuals who fit it can be tested and the distribution of their scores divided into hundredths. The result is a set of

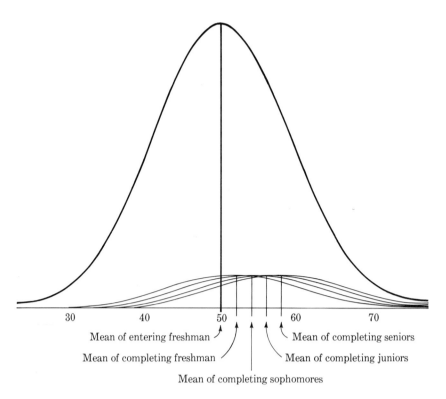

Figure 6-4   Distribution of four survivor groups artificially equated in size and superimposed upon original freshman distribution.

Table 6-1 Norms for Numerical Test.

Standard score	Percentiles				
	Entering freshman	Freshman	Sophomore	Junior	Senior
88					
86					
84					
82					
80					99
78				99	98
76			99	98	97
74		99	98	97	95
72	99	98	97	95	92
70	98	96	95	92	88
68	96	94	92	88	83
66	95	92	88	84	78
64	92	88	84	78	71
62	88	84	78	71	63
60	84	79	71	63	55
58	79	72	63	55	47
56	73	64	56	48	39
54	66	57	48	39	31
52	58	49	40	31	24
50	50	41	32	24	18
48	42	33	25	18	13
46	34	26	19	13	9
44	27	20	13	9	6
42	21	14	10	6	4
40	16	10	7	4	2
38	12	7	4	2	1
36	8	5	3	1	
34	5	3	2		
32	4	2	1		
30	2	1			
28	1				
26					
24					
22					
20					

norms like that in Table 6-1; now our Joe can see how his performance compares not only to the entering freshmen on whom the numerical test was originally standardized, but to any other group that has been tested. His score is better than 84 percent of entering freshmen, 79 percent of students completing their freshman year, 71 percent of completing sophomores, 63 percent of completing juniors, and 55 percent of graduating seniors. Those norm groups are students from all kinds of programs, including art, music, literature, and so forth, in which little numerical work is done. If Joe were thinking of entering an engineering curriculum, it would be appropriate to use norms derived from the testing of engineering students. Meanwhile, we should suspect that his score, compared to scores of engineering freshmen, might be well below the median and that among graduating seniors it might even be near the bottom. The small curves in Figure 6-4 have been artificially equated in size and shape. You might expect the "completing freshman" group to be closer to the size of the "entering freshman" group than the drawing would lead you to believe, and you would be right. You might also expect that the shapes of the distributions would change systematically from entering freshman to graduating senior; after all, it is the less able whose elimination accounts for the rising averages. But there you would be wrong. In practice, it turns out that distributions of more "select" groups tend to be very nearly normal.

One thing more. In Chapter 4 (Figure 4-5) we identified the proportions of the total sample that, in a normal distribution, are subtended by successive standard-deviation units marked off from the mean. Now, percentiles might be thought of as "cumulative percentages." Figure 6-5 reproduces 4-5, and adds the percentile that corresponds to each of the standard scores.

You should be able to reproduce this diagram from memory. But when I say "from memory," I do *not* mean rote memory. The only numbers you must memorize are "34" and "14." Once you have placed those properly, the rest are determined. Try it.

Age and Grade Norms

A percentile tells us where an individual is in relation to a specified norm group. An alternative way to interpret a scaled score

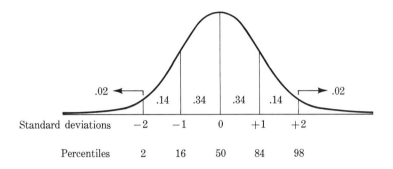

Figure 6-5 Normal distribution with base line divided into standard deviation units, showing proportion of total area subtended by each standard deviation, and listing cumulative percentages (percentiles).

(or sometimes even a raw score) is to find that group which the individual's score most nearly resembles. More specifically, each comparison group can be taken entirely from a particular chronological age group or from a particular grade in school.

By far the best known of age norms is the *mental age* (MA), which used to be a prerequisite to the derivation of an intelligence quotient (IQ). An individual child's score is compared to the average* scores of many chronological age groups; the age of the group whose average is closest to his determines his mental age. For example, if a six-year-old does as well as on intelligence test as the *average* six-year-old, his mental age is 6;* but if he does as well as the average eight-year-old does, his "MA" is 8, not 6.*

It has been more than a decade since standard (scaled) scores virtually took over the field of intelligence testing.* Now, the "comparison group" kind of interpretation is more often manifested in *scholastic achievement* testing as *grade*, rather than age, norms. If a child's score is nearest to that made by the average entering sixth grader, his *grade level* is said to be 6.1, or "the first month of the sixth grade." If the child's score were equal to the average pupil in the middle of his sixth year of school, his grade level would be 6.5. If his performance were most similar to the average seventh grader in his third month, his grade level would be 7.3; and so on.

Summary

Measurements in the social sciences are nearly always of *individual differences.* The report of a score can seldom be understood out of context; that is, it cannot be understood except in relation to other scores.

A *standard score* results from placing a score from any test onto a scale common to all tests. It is accomplished by dividing each individual deviation score by the standard deviation of its own distribution.

A *centile* (or *percentile*) shows where an individual stands in relation to a specified norm group. It does so by reporting just what percentage of that distribution falls below his score.

An *age* or *grade* level consists of the identification of the norm group whose average score is most similar to that of the individual being tested.

The essence of each of those procedures is *comparison* of an individual to a more or less well-defined group. A standard scale divides the base line of a distribution into equal parts, and percentiles divide the area under the curve (the total sample) into equal parts. Age or grade norms set their points of reference at the average scores of various age or norm groups. So all of these measures are different; but in every one, a new subject's score is compared to many others that have preceded it.

SAMPLE APPLICATIONS
FOR CHAPTER 6

EDUCATION

A fourth-grade student in your school has been having considerable difficulty in several subjects. His teacher has tried various approaches; but while the student can read words at grade level, he does not seem to remember what he reads, and although he can complete simple addition and subtraction problems, he cannot complete problems that involve several steps. The teacher refers the student to you, the school psychologist, and you give the student a number of tests to determine his specific strengths and disabilities. After the tests have been scored, what should you do before making your report?

POLITICAL SCIENCE

Many of the concepts in political theories are so complex that they cannot be conceived in terms of simple indices. As a result, researchers often create complex indices by summing several separately derived scores, each of which represents a different dimension of the concept. Imagine, for example, that you want to measure civil strife. A simple index (such as the frequency of riots) would not suffice for measuring such a complex concept; but adding the scores on man-hours lost per strike, assassinations per 1,000 population, and other such indices derived through different measurement procedures would not work either because each is measured in terms of a unit different from that used in measuring each of the others. What can you do?

PSYCHOLOGY

June is a year-old infant, the product of a difficult pregnancy and premature delivery. At birth, the attending physician feared that June was destined to show delayed or fragmented (high in some areas, low in others) development.

You are asked to evaluate the quality of June's overall development. You are concerned about her growth in three different areas: (1) intellectual; (2) social; (3) psychomotor. You administer separate tests corresponding to those areas and get three measures (scores). However, each of the three distributions has a different mean and standard deviation from the other two. How can you compare June's development in the three areas?

SOCIAL WORK

The State Personnel Office conducts an examination to establish a hiring register for social workers. It consists of a battery of tests concerning such functions as client assessment, client treatment, and community resources. Each test has its own mean and standard deviation, but all have been standardized on the same population. How can the Personnel Office inform you about your relative performance on the various parts of the examination?

SOCIOLOGY

Your data from the sociology exercise on authoritarianism in Chapter 5 include the score of your English professor for the coming semester. You fear that if he is very non-authoritarian his view may affect your relationship with him in class, as you are a red-necked conservative and highly authoritarian. How can you check this out?

7 | PRECISION OF MEASUREMENT

When Joe Cawledge makes a verbal test score of 30, we report his raw score, his standard score, or one or more percentiles. If someone wanted to know the mean score of a group of sophomores in a psychology experiment, we would have the same options there. Of course, neither *raw* score by itself would have any meaning; but what about the other two (standard score and percentile)? You learned in Chapter 6 to interpret *them*, did you not?

Well, yes and no. Back in Chapter 3 we discussed briefly the problem of "generalizing from a sample": whether we are justified, once we have measured a sample, in making statements about some population that we say the sample represents. Really, though, that is not an "either-or" question, and a proper answer would have to be phrased in terms of the "extent to which we are justified" in making such statements or of the "level of confidence" we have in making them. The probable *error* of such an inference is often as important to know as the inference itself.

This chapter then, is all about errors. Its title is not misleading, however, for it is by discovering the probable limits of error that we define the *precision* of our inferences. In many applications, that definition is actually more important than the score.

Standard Errors

A *statistic* is a statement about a sample. A mean is a statistic; so is a standard deviation. Together, they describe the sample.

But we are seldom interested primarily in the description of a sample. Usually, what we really want is a description of the *population* from which the sample has been drawn. Unfortunately, it is hardly ever possible to describe the population directly. (If it were, the sample and its population would be one and the same.) What we have to do is *infer* the characteristics of the population from those of the sample. The characteristics of the population are known as *parameters*. "Statistic" and "parameter" are parallel terms, one referring to the sample, the other to the population.[1] If I compute the mean height of 100 college men, I have a statistic; the corresponding parameter is the mean of *all* college men, most of whom I have not measured at all.

Example: The Standard Error of the Mean (SE $_{\overline{x}}$)

Let's take another example and work it through in more detail.[2] Suppose we want to know the mean IQ of college students. There are too many for us to measure all of them, but if it *were* possible, the distribution might look like the larger one in Figure 7-1, which is essentially the same as Figure 2-1.

But a second distribution is also depicted in Figure 7-1. It is the sample that we have drawn from the population. It is somewhat out of scale, for we should probably never be able to measure that large a proportion of the total population; but the drawing will suit our present purpose very nicely. The point is, of course, that the sample mean $(\overline{X})$ is not always the same as the population mean (μ_x).

Now comes the fun part. So far, we have only one sample—and indeed that is precisely what we would have in a "real life" situation. But this is not real life; this is an attempt to understand the *structure* underlying events that occur in the real world. To enhance that understanding, we shall have to imagine something that doesn't actually occur—namely, that after measuring one ran-

---

1. See also pages 13 and 17. The present chapter and the ones that follow it are concerned with "sampling" or "inferential" statistics, as distinguished from the "descriptive statistics" of Chapters 2 through 6.

2. The basic ideas in this section were introduced earlier in our discussion of the stability of the mean (pages 15–17). There they were presented in a most rudimentary way; now we can afford to be more precise, because since then you have mastered the concept of the standard deviation, which, as you will see in a moment, is essential to an understanding of standard error.

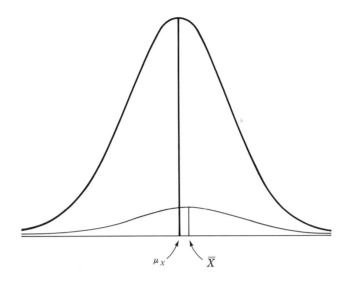

Figure 7-1 Distributions of IQs of college students—population and sample.

dom sample, we "throw it back," so to speak, and take another—and another and another, *ad infinitum*. If we were to do that, and if we were to obtain a statistic—say a mean—from each sample before we threw it back, we should eventually acquire a very large number of means, all of them drawn from the same population.

As noted above, those means are not all alike. As a matter of fact, if we plot them in a frequency distribution, we find that the distribution is normal. It is a distribution of *means;* like any distribution, it, too, has a mean (the mean of the.means). If we have, in our hypothetical investigation, taken an infinite number of samples, the mean of the means will be equal to the population mean.

So we have found that which we sought—pinned it down precisely! Unfortunately, however, it is only in fancy that we can take an infinite number of samples. In real life, we get only one; so how does all this really help? It helps because although a knowledge of the sample mean will never tell us what the population mean is, we *can estimate* from the variability and the size of the sample what the variability of a distribution of sample means would be. Our estimate of a standard deviation of a hypothetical distribution of sample means is called a *standard error* (SE$_{\bar{x}}$).[3]

$$SE_{\bar{x}} = \frac{S}{\sqrt{N}} \qquad (7\text{-}1)$$

Notice that the standard error is *small* when the sample is characterized by *small* variability in a *large* number of scores. A small standard error means that our sample is not very different from other samples that we might have taken from the population "collegiate IQs." It means that our obtained statistic is *reliable*.

Other Applications

Throughout the foregoing discussion of "standard errors," actually only one—the standard error of the mean—has been presented. That limitation was imposed because, although standard error may be a difficult concept to grasp initially, once it is understood in one situation it can readily be applied to others; the same logic applies to the standard error of a standard deviation, of a correlation coefficient, of an obtained score, and many others. In every case, a *small* standard error tells us that the sample "statistic" is a reliable estimate of the corresponding population "parameter."

Confidence Intervals and Levels of Confidence

We now have a powerful tool to use in our attempt to locate the population mean, because now we can try some hypotheses and discover the probabilities of their being valid. Figure 7-2 indicates graphically how that might be done in the case of the mean IQ of college students if the standard error of the mean were SE$_{\bar{x}}$ = .102. (Just suppose that it is .102. You don't have the data necessary to calculate it.)

---

3. Although it describes a hypothetical *population* (of sample means) rather than an actual sample, the standard error is regarded as a statistic rather than a parameter because it is estimated from the characteristics of a sample. It is an estimate of the standard deviation of a population of sample means ($\sigma_{\bar{x}}$). Since there is no limit to the size of a distribution of sample means, it is in principle impossible actually to measure any of its characteristics (parameters).

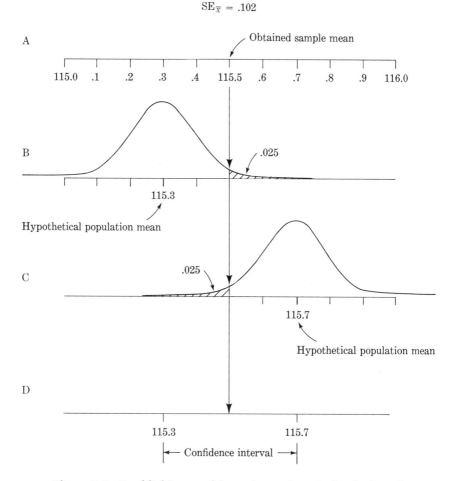

$$\mathrm{SE}_{\bar{x}} = .102$$

Figure 7-2   Establishing confidence interval at .95 level of confidence.

Essence of the Concept

The obtained mean IQ is 115.5. Figure 7-2 shows how that statistic plus the standard error of the mean can be used to answer the following question: "Within what limits may we be reasonably sure that the population mean resides?" The interval enclosed by those limits is called a *confidence interval,* and the "reasonably sure" in the above question is expressed quantitatively as a *level of confidence.*

You will recall that there is a variance in the means of random samples from the same population. That being the case, if the population mean were, say, .2 points *below* the obtained sample mean (Figure 7-2B), how many of a hundred sample means would be as *high* as the one we got? In the drawing, you can see that .025, or 25 out of every thousand, sample means would be that high. We can now state that the population mean is not lower than 115.3, and we can state it with a level of confidence of 1.000 − .025 = .975.

The lower limit is only half of what we need to define a "confidence interval"; but with that behind us, the rest is easy. All we do is to repeat the above process, except this time we will test the hypothesis that the population mean is *above* the obtained mean. Now we ask, "If the population mean were .2 points *above* the obtained sample mean (Figure 7-2C), what is the probability that a sample mean as *small* as ours would occur?" Again the probability is .025.

The testing of our first hypothesis informed us that there is a probability of only .025 that the obtained mean is lower than 115.3; the second test yielded a probability of .025 that it is higher than 115.7. Now let us put the two hypotheses together: "What is the probability that the population mean lies outside the interval 115.3 − 115.7?" The answer is, of course, .025 + .025, or .05; so if we say that the population mean *is* somewhere within that "confidence interval," we can do so at a "confidence level" of 1.00 − .05 = .95.

Some Exercises to Consolidate the Concept

The amount of deviation of an obtained mean from a hypothetical population mean is arbitrary; it depends on where you imagine that the true mean might be. But the shape of the distribution of sample means in Figure 7-2 is *not* arbitrary; it is estimated from sample size and variability by computing the standard error of the mean.

The deviation of an obtained mean from a population mean is a deviation score; $SE_{\bar{x}}$ is essentially a standard deviation. A deviation score divided by a standard deviation is a standard score. So when we divide our imagined deviation by our hypothetical standard deviation, we obtain a standard score. In Figure 7-2B, the "deviation score" is .2 and the "standard deviation" is .102; the resulting standard score is $.2/.102 = 1.96$. If we were to enter a "normal curve table" with a standard score of 1.96, we would find that the "area in the smaller portion" of the curve is indeed .025. If you were learning to *do* statistics, there would be such a table at the back of this book, and you would get much practice in using it. However, since your objective is to *understand* statistics, it is better that you deal with such problems in graphical terms. The following exercises will give you several opportunities to do that. Don't skip over them. On the other hand, don't be too concerned about precision; just do what *looks* right. That will be close enough for the purpose at hand; but if you think you need help in estimating areas under a normal curve, turn back to Figure 6-3.

Trace the curve in Figure 7-2 on a piece of paper, take a pair of scissors, and cut around it so that you have a distribution that can be moved about in order to test various hypotheses. Try the limits of 115.4–115.6, 115.2–115.8, 115.1–115.9, and 115.0–116.0; estimate in each case the probability first that the obtained mean is *outside* the interval, then that it is inside. You can do that for any interval you wish; but in a real-life situation, you would decide in advance how much risk of error you were willing to accept and conversely what confidence level you would require. Then you would find the score limits that corresponded to that level of confidence.

In other words, you would do just what I did in Figure 7-2. I decided initially that I wanted to illustrate a confidence level of .95, which leaves a probability of .05 that the obtained mean lies *outside* the confidence interval. Half of that .05 (i.e., .025) would occur above and half below that interval, so I proceeded as follows:
1. First, I slid the hypothetical distribution of means downward from 115.5 until the vertical line from the obtained mean cut off what appeared to me to be just .025 of the distribution. Then I designated the mean of that distribution as the lower limit of my confidence interval; if the population mean were any lower than that, less than .025 of repeated samples could be expected to have means as high as the one in my sample.
2. Having established the lower limit, I repeated the procedure in order to find the upper. I ask how high the distribution of sample means would have to be to put only .025 of them below the obtained mean of my sample.
3. Having found that, I had both limits of the confidence interval that corresponds to a confidence level of .95.

Effect of N on Standard Error

We have already noted that the variability of a hypothetical distribution of means $(SE_{\bar{x}})$ is estimated from the variability of an actual distribution of individual cases (S). But another factor is extremely important in making that estimate: the *number* of individual cases (N) in the sample.

The relation of N to $SE_{\bar{x}}$ may be less comprehensible initially than that of S to $SE_{\bar{x}}$, since the latter relation is between two forms of the same concept; but the relation to N can be quickly illuminated by means of a few simple diagrams. The top drawing in Figure 7-3 represents the actual distribution of an entire population; each "case" in the distribution is an individual member of that population. The three drawings below it are hypothetical distributions; each case within each of them is the mean of a sample from the above population of individuals. The differences among the three distributions are striking—and those differences are a function of the differences in N.[4]

You may have been surprised to see that the first of the hypothetical distributions of means was exactly the same as the original distribution of individual subjects. But a moment's reflection will convince you that it could be no other way, because when $N = 1$, every sample *is* an individual subject. Similarly, you can see that if every sample included the entire population, the variability among sample means would have to be precisely zero, and the statistic $(\bar{X})$ would be completely reliable. The middle drawing represents an intermediate situation in which every sample has an N of 25. In every case, the selection of samples is random, and each sample is returned to the population before the next one is taken.

Expectancy Tables

We have been describing ways in which sampling errors might be identified and quantified. We have done so in terms of the probability that some statement we make about a population is correct—i.e. that it is *reliable*.

---

4. The height of a drawing of a hypothetical distribution is arbitrary because, in such a distribution, N is infinite.

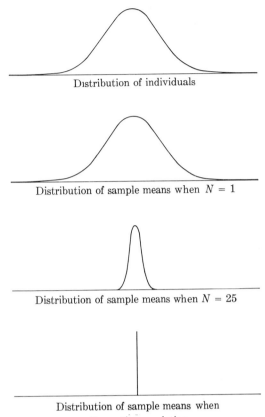

Distribution of individuals

Distribution of sample means when $N = 1$

Distribution of sample means when $N = 25$

Distribution of sample means when
N = entire population

Figure 7-3 Distributions of individuals and of the means of samples of three different sizes.

But there are many occasions in contemporary social science research in which a relation between *two* variables is the focus of interest. On such occasions, various correlation techniques may be applied. When a correlation coefficient is used to predict scores on one variable from scores on another, the accuracy of the prediction is known as *predictive validity*.

The Expectancy Table as a Scatterplot
The correlation coefficient can be a very useful statistic in those situations. However, if you need to communicate "relational" information to a person untrained in statistics (for

instance, a high school student or his parent) some other way must be found—some display that is concrete enough to be comprehended without benefit of previous study. A "scatterplot" (see Chapter 5) is one such display. You will recall that in seeking the Pearson r we were really attempting to define the slope of the regression line with both X and Y scores in standard units. The formula for r converts the scores into standard units and makes possible a very concise communication (r) *to anyone who is familiar with the process.* But by using a lot more space, we can communicate essentially the same information without converting to standard scores. A scatterplot displays that information, and most scatterplots use raw scores.

Figure 7-4 shows the regression of GPAs on test scores. The line across the drawing is the regression line—i.e., the *line of best fit.* (Notice that its slope is not .37, as it would have been if the two scales had been laid out in standard units.) It would be easy to predict GPAs from test scores by simply referring to that line. The only trouble with such predictions is that they leave out some important information. They ignore *variability.* If X were an aptitude test score and Y were grade point average at Cerebral University, the regression line would tell a student with a score of 57, for example, that in college his GPA would be 2.2. The entire scatterplot, on the other hand, would say that although a 2.2 GPA might be the central tendency of students who scored 57 on the test, it would not be the only possibility.

An *expectancy table* functions in the same way but is easier to use than a scatterplot. The space in the table is divided on two dimensions into rows and columns, thus forming an array of cells. If the entry in each cell is a frequency, then the scatterplot is almost exactly duplicated, as in Table 7-1. Usually, however, the frequencies in each column are transformed into percentages of the column sum, as in Table 7-2. In either case, you enter the table with the raw score from a student's test and emerge with more than one possible GPA.

Using an Expectancy Table

Now you can tell your student not only his most likely GPA in college, but also his chances of doing better (or worse) than the most probable performance. Joe Cawledge's score on Test X is 57; Table 7-2 says that of all students who previously scored in the 50's and subsequently have attended Cerebral University, 6 per-

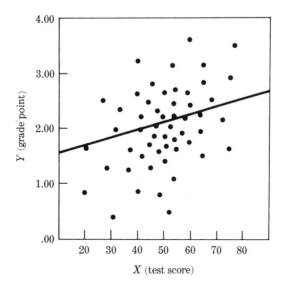

Figure 7-4 Scatterplot with superimposed regression line when $r = .37$.

cent have earned a GPA below 1.00, 41 percent have been in the interval 1.00 through 1.99, another 41 percent between 2.00 and 2.99, and 12 percent have achieved a 3.00 or better. That kind of information would be comprehensible to the interested layman. Indeed, it may be helpful even to the experienced tester or counselor; for although he knows immediately that any r below a magnitude of 1.00 implies some scatter, it is often difficult for him to visualize the *amount* of it. The expectancy table gets right down to cases.

Table 7-1 Expectancy of Frequencies (Figure 7-6; $r = .37$).

	10–19	20–29	30–39	40–49	50–59	60–69	70–79	80–89
3.00–3.99				1	2	1	1	
2.00–2.99		1	2	7	7	3	2	
1.00–1.99		2	3	7	7	2	1	
.00– .99		1	1	2	1			

Y

X

Table 7-2   Expectancy of Percentages (Figure 7-4; $r = .37$).

Y		10–19	20–29	30–39	40–49	50–59	60–69	70–79	80–89
3.00–3.99		0	0	6	12	17	25		
2.00–2.99		25	33	41	41	50	50		
1.00–1.99		50	50	41	41	33	25		
.00– .99		25	17	12	6	0	0		

X

A refinement often made in the interpretation of such a table is the derivation of *cumulative* percentages. Table 7-3, for example, implies not that Joe Cawledge has a 41 percent chance of achieving a GPA between 2.00 and 2.99, but that the probability of his doing *2.00 or better* is .41 + .12 = .53. That sum is often more useful than either probability by itself—so often, in fact, that many expectancy tables are made up entirely of cumulative percentages. Table 7-3 is based upon the same data as 7-2; the only difference is that all the percentages are cumulative. That makes it less flexible than Table 7-2; but, for the many persons who are interested primarily in cumulative percentages, it is easier to use.

Summary

Every measure is derived from a sample, and although samples are, by definition, representative of the population from which they are drawn, they differ from each other even when drawn from the same population. If there are no *systematic* deviations of the characteristics of the sample from those of the parent population, we say that the deviations are *random,* and we refer to them collectively as "error" variance.

Now, if there is going to be error in our measurement, it behooves us to know its magnitude—or rather, its *probable* magnitude—for we might otherwise be over-confident of the statements that we make about the population from which any sample

Table 7-3   Expectancy of Cumulative Percentages (Figure 7-4; r = .37).

Y		10–19	20–29	30–39	40–49	50–59	60–69	70–79	80–89
3.00–		0	0	6	12	17	25		
2.00–		25	33	47	53	67	75		
1.00–		75	83	88	94	100	100		
.00–		100	100	100	100	100	100		

X

has been drawn. The most common way of quantifying error is to compute a *standard error* for whatever statistic we wish to cite.

A standard error is an estimate of the standard deviation of a hypothetical distribution of values that would be obtained for a given statatistic if repeated samples were drawn from a single population. If the statistic in question were a mean, for example, we could estimate the variability of a distribution of means of successive samples of the same size from the same population. The standard deviation of that distribution is estimated by the *standard error of the mean.*

The standard error of the mean is important because with it we can mark off the *limits* within which the population mean probably lies, and we can ascertain what that probability is. In technical terms, we can state the *level of confidence* of our hypothesis that the population mean resides within the specified *confidence interval.*

The size of the standard error of the mean is a function of the variability of the sample, as indicated above. But it is also a function of the *number of individuals* in the sample. N varies all the way from *one* individual at one extreme to *all the individuals in the population* at the other. If each sample includes only one individual, the standard error is as large as the standard deviation of the population would be if we could get it. If the sample size is the same as that of the population, the standard error is zero, because repeated samples would all have the same mean. Other sample sizes have intermediate effects on the standard error.

The standard error of the mean has been used here as an illustration of the error concept. Everything that has been said about it can be said also of the standard errors of other statistics.

Finally, there is another way to deal with "error" in situations that call for making predictions of an individual's position on one variable from a knowledge of his position on another. An *expectancy table* can inform a person who has a given score on the predictor variable (say, a scholastic aptitude test) what his chances are of achieving each of many possible scores on the criterion measure (say, grade-point average in college). The table thus deals explicitly with that which a correlation coefficient only implies—namely, the deviation of criterion measures from their regression line.

SAMPLE APPLICATIONS
FOR CHAPTER 7

EDUCATION

You are superintendent of a large urban school district. You want to know the mean achievement levels of elementary-school children in each grade. Funds are limited, so you test a random sample of 10 percent of the students at each grade level, using a nationally standardized achievement test battery. From that information you calculate the mean achievement level of the sample at each grade level. What else might you want to know besides the mean?

POLITICAL SCIENCE

You have developed an Index of Political Participation and as a part of the standardization procedure have applied it to each of several middle-class neighborhoods. You report the mean of the scores (50 points) to potential users of the Index, but since no measure is perfectly reliable you want to report also the limits within which the *true* mean probably lies. How do you proceed?

PSYCHOLOGY

You have administered an inkblot test to a ten-year-old boy. The results indicate that the boy is mildly disturbed and in need of psychotherapy. Since you know, however, that the inkblot measure is not perfectly reliable (i.e., the results may vary from one administration to the next), you wonder whether psychotherapy should be recommended. The next administration of the inkblot test might indicate that the child is well within normal limits. The test manual might contain something that would help you to decide how much confidence to place in the test score. What would you look for in the manual?

SOCIAL WORK

A family service agency utilizes a Family Life Involvement Profile to measure a family's psychological functioning. The scale yields a score that indicates whether a family's functioning is inadequate, marginal, or adequate. In your assessment of a family, you obtain a FLIP score that indicates a marginally functioning family. But you wonder how accurate the score is—how much confidence you can have in the rating. How can you quantify that uncertainty?

SOCIOLOGY

A member of your research class notes that you have only a *sample* of professors in your project on authoritarianism (sociology exercise, Chapter 5). He asserts that since you have only a sample, you do not really know where the true population mean is, and that consequently our conclusions are meaningless. How do you respond?

8 | SIGNIFICANCE OF A DIFFERENCE BETWEEN TWO MEANS

Often, in both basic and applied research, it is important to know whether two populations are different from each other. From the point of view of the researcher, the question might better be stated, "Do the two samples that I have measured represent two populations, or are they merely two samples from the same population?" The true answer to that question is either "yes" or "no." The researcher can never know the true answer, however; so it must be stated in probabilistic terms. This answer depends on two factors: (1) the size and direction of the obtained difference between the means and (2) the variability of a hypothetical distribution of differences that have been obtained in the same way that his was.

One way to illustrate the effect of variability (sampling error) on his answer is to analyze carefully a single example. In the following section, we shall examine an experimental study in the field of education, from the design of the experiment to the announcement of the results, concentrating throughout on the basic ideas rather than the computations.

An Example

Imagine that we have invented a new—and we hope better—method of teaching French grammar to American high school students who have no knowledge of that language. Is it really better than the prevailing method? To find out, we take two samples from the population "Naive American High School Students," make sure that the two are initially random samples from a single population, and teach one group (hereafter to be known as the control group) by the traditional method and the other (the *experimental group*) by the new method. Then, after 150 hours of teaching time, we test both groups and compare their mean scores. The objective of our study is to ascertain whether the two are *still* "random samples of the same population" insofar as their knowledge of French grammar is concerned.

Imagine that the difference between the means is 10 points. Is the obtained difference significant? Is it large enough that we may reject the hypothesis that the control and experimental groups remain, after the teaching as they had been before, random samples from a single population with respect to knowledge of French grammar? That single population might be labeled "American High School Students Who Have Had 150 Hours of Instruction in French Grammar." Our hope is that, instead, there are *two* populations—one superior to the other with respect to knowledge of French grammar—and that the experimental group represents the superior population. This is admittedly very abstract, because you have to imagine a population that does not at the moment exist—that is, a population made up of an infinite number of American high school students who have been taught French grammar by the new method. We suspect that such a population would be superior to the one with which we are more familiar. But before we can be sure of that, we must disprove the hypothesis that the two groups are merely two samples of a *single* population. That is the hypothesis of no difference, or the *null hypothesis*, and our test of significance is really a test of the null hypothesis: we attempt to disprove the hypothesis that there is no real difference between the groups—or rather, that the observed difference is merely a chance difference resulting from ordinary sampling error.

That may sound easy, but there is a complication. Like virtually every conclusion that emanates from statistical reasoning, this one must be stated not in absolute terms, but in terms of probability. We will not get a "yes" or a "no" out of our statistical test. Rather, it will give us the probability that we are rejecting a *true* null hypothesis, and although that probability could conceivably approach zero, we'll never get a flat-out "no."[1]

---

1. If we do *not* reject the null hypothesis, we must regard it as tenable (our data offer insufficient evidence against it) but not necessarily as true. Other hypotheses are also tenable in this sense (for instance, the true difference *could* be identical to the one we obtained from our samples).

We therefore adopt arbitrarily some level of probability that is *an acceptable approximation* of zero, and when a test reveals a probability below that level, we reject the null hypothesis. (After that, we may entertain others.) Whether a particular difference—like the 10 points we obtained between our control and experimental groups after the language instruction—meets that preset criterion depends upon *the variability of a hypothetical distribution* in much the same way that the size of the "confidence interval" did when we were discussing the stability of the mean. It might be worth your while to review that section before proceeding with this one; it begins on page 80.

In fact, a knowledge of the standard errors of the means of both groups is essential to our present purpose. This time, however, we are dealing primarily with a hypothetical distribution of *differences between* means—in this case, between the means of groups of American high school students taught by two different methods.

Test of Significance: The z Ratio

Is a difference of 10 points sufficiently large to be significant, or is it small enough that it might easily have occurred by chance—by ordinary sampling error? The answer is a function not only of the size of the difference itself but, very importantly, of the sampling errors—as indicated by the standard errors of the means—of the two groups being tested. The next six paragraphs, together with Figures 8-1 through 8-4, concern a situation in which the standard error of the mean is rather small and is the same for both groups. Just accept my figures for the standard errors in Figures 8-1 through 8-8. Concentrate now on what you can *do* with a standard error once you have it.

Look at Figure 8-1. It shows two hypothetical distributions of sample means on a scale of French grammar scores; the solid line represents control groups, the dotted line "experimentals." They occupy the same space because the drawing was made on the hypothesis that the two kinds of groups are in reality *not* "two kinds" with respect to achievement in French grammar, but are merely two sets of random samples from a single population. That is, of course, the "null hypothesis," and we shall try our best to disprove it (or rather, to make it untenable).

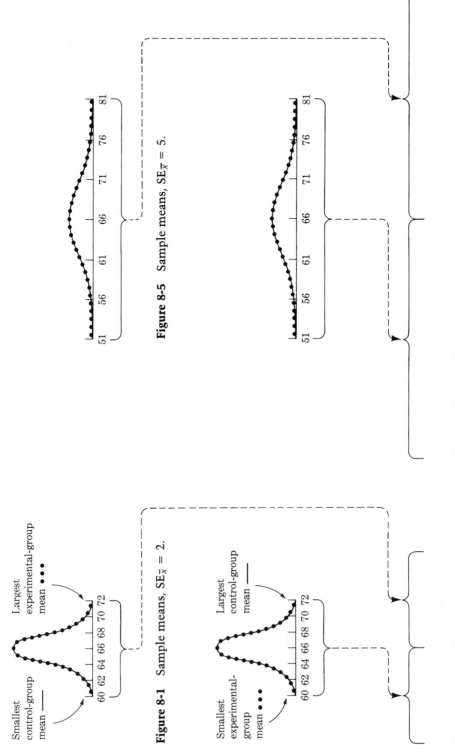

Figure 8-5 Sample means, $SE_{\bar{x}} = 5$.

Figure 8-6 Sample means.

Figure 8-1 Sample means, $SE_{\bar{x}} = 2$.

Figure 8-2 Sample means.

Figure 8-3 Differences between means, $SE_{\bar{x}_e - \bar{x}_c} = 2.83$.

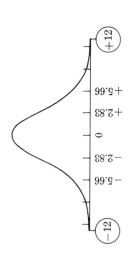

Obtained difference = 10

Figure 8-4 Differences between means.

Note: Figures 8-1 through 8-4 show test of significance of differences when $SE_{\bar{x}_c}$ and $SE_{\bar{x}_e}$ are small. Distributions are all hypothetical.

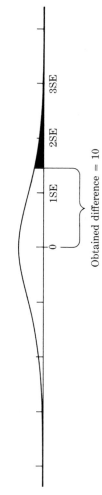

Figure 8-7 Differences between means, $SE_{\bar{x}_e - \bar{x}_c} = 7.07$.

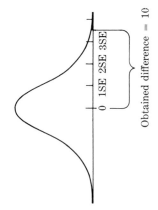

Obtained difference = 10

Figure 8-8 Differences between means.

Note: Figures 8-5 through 8-8 show test of significance of differences when $SE_{\bar{x}_c}$ and $SE_{\bar{x}_e}$ are large. Distributions are all hypothetical.

[99]

In order to test the null hypothesis, it is necessary to think in terms of a scale not of scores but of *differences between scores*—specifically, of differences between control- and experimental-group means. Look again at Figure 8-1 and imagine taking pairs of means (one control and one experimental group in each pair) at random from the two distributions. Most of the intra-pair differences would be close to zero, but a few—just by chance, remember—would be substantially larger. In fact, if you were to compare the largest experimental-group mean with the smallest mean of the "controls," the difference would be 12 raw-score points, or 6 standard errors.

That difference, $\overline{X}_e - \overline{X}_c$, is in the direction we have predicted, so we'll call it "positive." But if all the samples are drawn from the same population, as the null hypothesis would have it, there should be as many "negative" differences as there are positive, and the largest of them should be just as impressive as the positive difference we just found. Figure 8-2 represents the same two distributions as Figure 8-1; if you will compare the *lowest* experimental with the *highest* control group mean, you will indeed discover that the difference is again 12, but this time in the "negative" direction.

Figures 8-1 and 8-2, then, represent the same pair of distributions analyzed in two different ways: one as positive, the other as negative differences. Figure 8-3 combines "positive" and "negative" differences into a distribution of differences. Positives are on the right, negatives are on the left, and the mean is zero—the null hypothesis again. If we were to compute an infinite number of differences between pairs of means, selecting each pair at random from a single population, that is about what the distribution would look like.

With that much behind us, the testing of the null hypothesis should be relatively easy. All we have to do is place our obtained difference within the hypothetical distribution of differences, as in Figure 8-4, and we can see immediately that only a very few positive differences[2] that large occur simply by chance among samples taken from a single population.

---

2. Asking only about *positive* differences constitutes a "one-tail test." A discussion of one- vs. two-tail tests begins on page 104.

By inspection it appears that hardly any of such differences would be as large as the one we obtained in our experiment; we are justified, therefore, in rejecting the null hypothesis as untenable. Formula 8-1 offers a more precise estimate of the variability of differences between means.*

$$SE_{\bar{X}_e - \bar{X}_c} = \sqrt{SE_{\bar{X}_c}^2 + SE_{\bar{X}_e}^2} \tag{8-1}$$

where $SE_{\bar{X}_e - \bar{X}_c}$ is the standard error of a difference between means, $SE_{\bar{X}_c}$ is the standard error of the mean for the control group, and $SE_{\bar{X}_e}$ is the standard error of the mean for the experimental group. Since our interest is in the underlying logic of the significance test rather than in the precision afforded by the formula, we shall pass by the latter with only the comment that, like Figures 8-1 through 8-4, the formula shows that the variability of random *differences between* means taken in pairs ($SE_{\bar{X}_e - \bar{X}_c}$) is proportional to variability among means taken singly ($SE_{\bar{X}_c}$ and $SE_{\bar{X}_e}$).

To illustrate that point further, and to emphasize the importance of $SE_{\bar{X}}$ in determining the fate of the null hypothesis, turn back to page 98 and see how our 10-point obtained difference would have fared had the sample means been less stable (Figures 8-5 through 8-8). Whereas in Figure 8-4 our difference was 3.53 standard errors ($z = 3.53$) above the mean and had a probability of .0002, in Figure 8-8, a difference of ten is only 1.41 standard errors away from the mean and has a probability of .0793. In the latter case, if we were to take 1000 pairs of samples at random from *a single* population, 79 of the intra-pair differences would be at least as large as ours. Had we obtained our difference in those circumstances, we would have had to accept the null hypothesis as tenable. Incidentally, do not be disturbed if you cannot estimate the above values with this kind of precision simply by looking at the graphs; I did it by formula.*

All of the examples in this section have been treated graphically in order to expose the structure of their operations. Normally, however, those operations are done algebraically for greater accuracy. In the language-teaching study, the formula would be

$$z = \frac{\bar{X}_e - \bar{X}_c) - 0}{SE_{\bar{X}_e - \bar{X}_c}} \tag{8-2}$$

where z is the ratio of an *obtained difference* to the *standard error of a difference* between means in a distribution of such differences when the mean of the distribution is zero. In Figure 8-4, that ratio is 10/2.83; in Figure 8-8, it is 10/7.07. The zero in the formula has no effect on the outcome; it is there only to remind you of precisely what the formula is supposed to do: namely, to ascertain how far our obtained difference is from the mean of a distribution of differences, all of which are the results of sampling errors. The mean of such a distribution would indeed be zero.

Test of Significance: The *t* Ratio

We may regard Formula 8-2 as the basic formula for the test of significance because it clearly reveals the relation between the difference and the standard error of the difference. When that difference is obtained from a small sample, however, it is not the z ratio that is tested for significance, but a ratio called "t." The t ratio is essentially the same as z, but it is referred for significance testing to a distribution that changes its shape (because the probability of obtaining the various scores in it changes) as N gets smaller. When N is larger than 30, the difference between z and t distributions is negligible; when N is infinitely large, the two are identical.[3] In Figure 8-9, it is clear that when the t distribution is used, the probabilities to be inferred from various placements on the base line are in many instances quite different if N is small than if it is large; most notably, when N is small, extremely large ratios (either positive or negative) make up a larger than normal part of the distribution.

Because it is appropriate for use with either large *or* small distributions, the t test is used almost universally in place of z. But because z is a logical extension of concepts developed earlier, it is best to think of t as a modified z.

---

3. You will notice that in Figure 8-9, sample size is indexed by "df" (degrees of freedom) instead of by N. That is really all you need to know in order to follow this discussion. However, the concept of "degrees of freedom" is discussed briefly in a note in the back of the book (the one referring to Chapter 4, page 25).

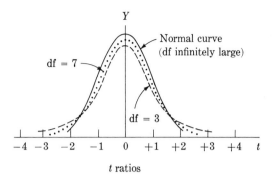

Figure 8-9 Distribution of *t* ratios at three different sample sizes. [Adapted from Henry L. Alder and Edward B. Roessler, *Introduction to Probability and Statistics*, 6th ed. W. H. Freeman and Company. Copyright © 1977.]

Significance Levels

We have seen above that in certain circumstances a difference of 10 points between two sample means might occur frequently by chance. However, our difference was obtained not in the circumstances depicted in Figures 8-5 through 8-8, but in those of Figures 8-1 through 8-4; there it is clear that the probability is extremely small that, with respect to their knowledge of French grammar, our control and experimental groups are random samples of a single population. Now we may announce to an eagerly waiting public that our new method of teaching French grammar is almost certainly better than the traditional one—at least under the circumstances prevailing in our experiment. We proclaim that superiority at a *significance level* of .0002, because there is only that much of a probability that our proclamation is wrong—that we are rejecting a true null hypothesis.

It may seem to you that the "level of significance" in this instance should be 1.000 − .0002, or .9998. If it does, I agree with you; but tradition holds that the level must be expressed as *the probability that a true null hypothesis is being rejected.* That means that the *lower* the significance level, the *higher* is our confidence that the effect we have observed is real—that it is *reliable.**

Tradition also provides us with two levels recognized as significant and very significant, respectively. A "significant" difference is less than .05, which is often written "$p \leq .05$." A "very significant" difference is one for which the probability of having occurred by sampling error is less than .01; it is sometimes written "$p \leq .01$." Finally, in tables and other places where the briefest possible abbreviations are used, a "significant" difference sometimes is designated simply by the letter "S" and a "very significant" one by "VS." In practice, those two levels are often used but also often ignored. For example, a researcher who had obtained as impressive a difference as that between our control and experimental groups in the language-testing study would be unlikely to hide his light under the bushel "less than .01"!

One- vs. Two-Tail Tests

One last thing should be said about the testing of significance and the levels of confidence that emanate therefrom. In footnote 2, I told you that we were doing what is called a *one-tail test* of the significance of our obtained difference of 10 points; we were saying that there was only a .0002 probability that a difference that large *and in the expected direction* would occur by chance. The alternative is a *two-tail test*, which is appropriate whenever *we have not predicted the direction* of the difference. If you will look back at Figure 8-3 for a moment, you will see that the probability of a difference of 10 in *either* direction is double that of a difference in *one* direction. That means that a particular difference conceivably could fail the significance test if there has been no directional prediction and pass it if a prediction has been made (provided, of course, that the outcome matches the prediction). But we are not free to select either test arbitrarily; if we have no reason to expect a difference in one direction rather than another, we are obligated to use the two-tail test. It is only when we do have such a reason, *when we can make a prediction in advance*, and when the results confirm the prediction, that we are justified in using the one-tail test. *

Statistical vs. Practical Significance

The length of this section barely exceeds that of a long footnote, but it contains an important notion that might easily be overlooked if it were not emphasized. The point is that even though a difference is shown to be statistically significant, it may not have any *practical* significance.

The result of our language-teaching experiment was very convincing; there can be almost no doubt that the difference between the two groups at the end of the teaching period was real. But how does that information affect the decisions of a school administrator who must decide whether to adopt the new method? That depends on many considerations, some of which have nothing to do with statistics. It depends to a great extent on how costly the new teaching method is to implement—whether it requires expensive equipment or specially trained teachers. The administrator's decision may also depend on the absolute size of the difference. That may seem a contradiction of everything we've been saying up until now; and it is, unless we keep in mind the distinction between statistical and practical significance.

The joker is in that standard-error term—more specifically, the importance of N in determining its size. If you analyze either Figures 8-1 to 8-8 or Formula 8-1, you find that the size of the standard error of the difference is proportional to the standard errors of the means of the two samples. And the size of a standard error of the mean is determined partly by the variability in the sample and partly by its *size:*

$$\mathrm{SE}_{\overline{X}} = \frac{S}{\sqrt{N}} \qquad (7\text{-}1)$$

Now, the significance of an obtained difference depends upon the ratio between that difference and the standard error of the difference:

$$\mathrm{SE}_{\overline{X}_e - \overline{X}_c} = \frac{\overline{X}_1 - \overline{X}_2}{\mathrm{SE}_{\overline{X}_1 - \overline{X}_2}} \qquad (8\text{-}3)$$

That means that if our samples are large enough, *any* difference will be statistically significant, no matter how small it may be; and while a z ratio of 3.53 indicates that the difference, no matter how small, is probably genuine, a school administrator might decide that a very *small* difference would not be worth its additional cost, even if it were *absolutely* reliable.

The size of a difference is in no way affected by its reliability. The temptation to infer that it is should be firmly suppressed.

Summary

Sometimes it is important to know whether two groups are different from each other—or rather, whether they *represent populations* that are different from each other. In this chapter, we have compared the performance of two groups of students after exposure to two different teaching methods. One group, taught by the traditional method, was called the *control group;* the other, taught by a new method, was the *experimental group.* The experimental group outscored the controls.

But two groups given the same treatment could have scored differently by chance, and in such a case we would have to admit that the obtained difference was a *sampling error*—that the two groups are really only two samples from the same population with respect to the performance being tested. We must consider the possibility that our control and experimental groups are like that, too—that the difference we obtained was due entirely to sampling error and that the two groups are really only two samples from a single population. That possibility is known as the hypothesis of no difference, or *null hypothesis.*

The general question, then, is really, "How large does an obtained difference have to be before we are justified in rejecting the null hypothesis?" The procedure by which that question is answered is known as a *test of significance.* In the case at hand, we ask a more specific form of the question, namely, "Is our obtained difference large enough to justify a rejection of the null hypothesis?"

The answer will not be a simple "yes" or "no"; it must be given in terms of *probability.* What is the probability that a difference as large as ours would occur if there were no real difference in the effectiveness of the two teaching methods? Acceptable levels of probability are somewhat arbitrary, but two such *levels of confidence* have traditionally been set at .05 and .01. However, a researcher who obtains a null hypothesis probability much lower than .01 will probably report it exactly, because the lower it is, the more impressive is the outcome of his study.

Is it possible to ask all of the above questions about any measurement of any two groups of subjects? We could use the test of significance simply to explore—to search out differences worthy of further investigation. But in the case of the two teaching methods described above, we were not exploring; we *expected* the experimental group to be better than the control. That is an important distinction, because if we can state our expectancy in advance, we are privileged to use a *one-tail test* instead of a *two-tail test* of significance. If we are exploring, we must ask, "What is the probability of a sampling error this large *in either direction?*"; whereas, if we have stated our expectancy, we may ask instead, "What is the probability of obtaining this large an error *in favor of the experimental group?*" Since the probability of a difference in *one specified* direction is just half that of the same amount of difference in *either* direction, the one-tail test is twice as sensitive as the one we would be obliged to use if we were merely exploring; a difference half as large will qualify at whatever level of confidence we have prescribed. Finally, a test of statistical significance tells us how reliable our difference is, but that is only one factor in the making of practical decisions.

EDUCATION

You are principal of a high school. The teachers, counselors, and administrators of the school have developed a one-semester group-counseling program for students who are disrupting class to help them learn more appropriate ways of resolving conflicts and participating in classroom activities. To find out whether the program helps reduce student disruptiveness, you randomly assign half of the fifty most disruptive students to the group-counseling program. At the end of the semester, the teachers rate the disruptiveness of all fifty students. When resulting scores have been collected, what do you do with them?

POLITICAL SCIENCE

Increasingly in recent years, political scientists have used statistical methods to evaluate the effectiveness of public programs. An example of this kind of research can be seen in the following case involving a crime control program. The citizens of Gritty City have demanded that local officials do something to resolve the problem of burglaries. You are chief of police. As a first step, you propose that the city council establish a special task force that will give presentations in each neighborhood on how to prevent burglary. Since the council is hesitant to fund the task force on a permanent basis without any evidence regarding its impact on burglaries, its members agree to make a decision after analyzing data from a pilot program. To set up such a program, you draw two random samples from the population comprised of all of the city's precincts. The residents in one group then receive the task force presentations; residents in the other group do not.

After three months, you compare the mean number of burglaries for the citizen-participation precincts with that of the cops-only precincts. What is an appropriate statistic for making such a comparison?

PSYCHOLOGY

As a clinical researcher, you are interested in ascertaining how a period of training in muscle relaxation will compare to the use of stimulant drugs in reducing hyperactivity in young children. You assign half of the children medically classified as hyperactive to the relaxation program and half to the drug program. Following a thirty-day period of intervention, the activity level of all children in the study will be assessed by means of a rating scale. There will almost certainly be some difference between the two groups. How can you tell whether the difference is significant?

SOCIAL WORK

As a social worker in a senior citizens' center you are concerned about the health and vitality of the seniors who frequent the center. It is your assessment that the center's current program of bingo, pool, backgammon, quilting, films, and occasional field trips is not sufficient to keep seniors active and alert; for they experience numerous health problems, including strokes, heart attacks, upper respiratory illnesses, and emotional illnesses involving depression and anxiety. After attending a workshop on services for the elderly, you plan to implement a new program which involves group discussion, meditation, and physical exercises of stretching and posturing. It is a structured program; seniors meet for two hours twice weekly. In order to ascertain the impact of this new program you select half of the seniors on a random basis and engage them in the program for a year. You then plan to compare this group to the half of the membership that has continued in the regular center activities. After collecting health data on all of your subjects, you find a difference in favor of the experimental group. How can you estimate the probability that this difference arose by chance?

SOCIOLOGY

A family-planning agency has asked you whether Catholic families are larger than non-Catholic families in your state. You draw a random sample from census data and find that the mean size of Catholic families is indeed larger than that of non-Catholic families. How might you test the significance of this difference?

9 | MORE ON THE TESTING OF HYPOTHESES

A chapter with this title could easily occupy as much space as all the rest of the book put together. It will not, because I have selected just two tests of significance as illustrations. The others will be mentioned only briefly, if at all. For our purposes, it is not necessary to describe those others. In fact, probably the most important contribution that I can make to your understanding of all of them can be stated without *any* illustrations. The fundamental idea is this: each of them is conceived as *the testing of a null hypothesis*—a hypothesis of no difference.*

The difference that emerged from our experiment with two methods of teaching French grammar was a difference in *amount.* We asked whether an innovative method produced students who were more able than those produced by the more traditional method with which it was being compared. More technically, we asked whether the difference we obtained was attributable to the difference in teaching methods or whether it was merely the result of sampling errors—that is, we tested the null hypothesis. But we could have asked a different question: given some clear criterion of "passing," does the innovative method produce fewer failures than the traditional? That question is posed in terms of *frequencies.* Indeed, frequencies are often used to indicate amounts, as when number of words correct indicates amount of typing skill or number of strokes (or rather its inverse) indicates amount of golfing skill. The significance tests described in Chapter 8 can be used in such cases, and they can be modified to deal with the categorical cases just mentioned (pass–fail, yes–no, innovative–traditional, etc.). More often, however, such cases are analyzed in a different way; so the first part of the present chapter will examine a test of the null hypothesis in which all of the data are in the form of frequencies.

Another feature of the example that we used in Chapter 8 is not characteristic of all experiments: only two "treatments" were compared. What should we do if we wanted to assess the relative effectiveness of several teaching methods? Or what if we suspected that one treatment method might be effective in the hands of one teacher and another might be superior when used by a different teacher? The second section of the chapter will describe a kind of analysis by which both of those questions may be answered.

Notice that every "test of significance" is a test of the null hypothesis. Conversely, any test of the null hypothesis is a test of significance; and since any observed difference could conceivably have resulted from sampling errors, any difference may be put to the test of significance.*

Comparison of Frequencies: Chi Square

In the introduction to this chapter, I mentioned that the scale of scores used earlier in our investigation of teaching methods could have been reduced to two scores only: passing and failing. I pointed out that under those circumstances, the data would probably be in the form of frequencies indicating how many subjects passed in each of the two conditions.

In practice, however, test-score scales are seldom reduced to two class intervals; to do so would be to throw away information. Significance tests designed to deal with frequencies, proportions, or probabilities are usually applied in situations that do not produce finely divided scales in the first place. In a public opinion poll, for example, respondents are usually asked a question to which they are expected to answer either "yes" or "no," "for" or "against," and so forth.[1]

Let us consider briefly the reasoning behind a significance test of a set of data from a public opinion poll. Imagine that a

---

1. There are ways of obtaining more finely graded responses, but they are more cumbersome to administer than the single question. Furthermore, a relatively crude index is often the most appropriate to the circumstances. For example, in predicting the outcome of an election, every mark that a voter makes on his ballot represents what is essentially a "yes-no" decision.

Table 9-1 Arrangement of Data in Election Problem.

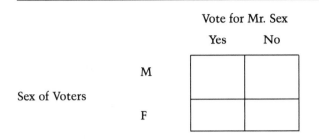

particular male candidate for public office has considerably more sex appeal than another but seems to differ very little from his only opponent (another male) on any other dimension. He wins, and we wonder whether his sex appeal has played a part in his victory.

If we assume that the voters of both sexes have a basically heterosexual orientation, we may get an informative answer by rephrasing the question: "Did women vote for Mr. Sex in significantly larger numbers than did men?" The data can be arranged in a 2 × 2 table (see Table 9-1). If there are in this election 20,000 voters—10,000 male and 10,000 female—of which we have a 1 percent representative sample, if Mr. Sex has received 60 percent of all the votes cast, and if there are *no* differences between men and women with respect to voting behavior in this election, the table will look like Table 9-2.

You should recognize that last "if" as the null hypothesis; our test of significance will be an attempt to show that it is unten-

Table 9-2 Expected Frequencies (Null Hypothesis).

		Votes for Mr. Sex		
		Yes	No	Total
	M	60	40	100
Sex of Voters	F	60	40	100
	Total	120	80	200

Table 9-3   Obtained Frequencies (From Actual Sample).

| | | Votes for Mr. Sex | | |
		Yes	No	Total
	M	50	50	100
Sex of Voters	F	70	30	100
	Total	120	80	200

able. To do so, we shall have to demonstrate (1) that our *obtained* frequencies differ from those that would be *expected* if there were no real difference between the voting behaviors of men and women and (2) that the difference is larger than we should be likely to get through sampling errors.

The division of votes in our sample turns out to be as in Table 9-3.

Now, in Table 9-4, we'll put expected and obtained frequencies together, the better to compare them. It is clear that our candidate received a higher proportion of the female vote than of the male. But is the difference *significant*? How likely is it that a difference (actually a set of four differences in Table 9-4) as large as that would occur in a random sample if there were no difference at all within the population of voters? (The null hypothesis here

Table 9-4   Obtained Minus Expected Frequencies.

| | | Votes for Mr. Sex | |
		Yes	No
	M	50 − 60 ───── − 10	50 − 40 ───── + 10
Sex of Voters	F	70 − 60 ───── + 10	30 − 40 ───── − 10

is that the two variables—Sex of voters and Votes for Mr. Sex—are *independent* of each other.)

To find out, we may use a statistic called *chi square* (χ^2). The general formula is

$$\chi^2 = \Sigma \frac{(f_o - f_e)^2}{f_e} \qquad (9\text{-}1)^*$$

where f_o = obtained frequency and f_e = expected frequency. Focus on one part of that formula for a moment

$$\frac{(f_o - f_e)}{f_e} \qquad (9\text{-}2)$$

and you will grasp intuitively the fundamental nature of chi square. The greater the deviation from the expected frequency (expressed as a proportion of that frequency), the larger is chi square; and since the expected frequency is "expected" on the hypothesis of no difference, chi square is an index of deviation from that hypothesis. Tables are available in which one can find the significance level of any given chi square. In the present example, chi square is 8.34, which is significant at the .01 level; the probability that our obtained difference was due to sampling errors is less than 1 percent.*

Multimean Comparisons: Analysis of Variance

In Chapter 8, we applied a significance test to a difference between two means. That is frequently needed in behavioral research, but it certainly is not always appropriate. Often an experimental design will call for more than two categories of the independent variable, and sometimes systematic changes are made in more than one independent variable.[2] Tests based on the

---

2. An *independent variable* is one that is manipulated by the experimenter. (Mathematically, it is independent; experimentally, it is really a *manipulated* variable. But the mathematical term is always used, whether the context is mathematical or experimental.) A *dependent variable* is one whose values are determined (in cases in which the null hypothesis is properly rejected) by those of the independent variable(s).

z and t ratios described in Chapter 8 cannot be used to assess significance in either of those situations. But one technique is appropriate to either or both of them; it is called *analysis of variance* (ANOVA).

One-Way Analysis of Variance

In this section, we shall be concerned only with the former of the two designs mentioned above—namely, the one in which (1) there is one independent variable, but (2) there are more than two categories of that variable. Our example will be a simple extension of the experiment we used to illustrate the application of the z (or t) ratio. There, we had two groups, which we called "control" and "experimental"; here, we shall have six groups, and we shall refer to them as simply I, II, III, IV, V, and VI. There, we were comparing a new method of teaching French grammar with a traditional method; here, there are six different methods, one or more of which may be traditional. Let us say that the control and experimental groups of the earlier experiment are Groups I and IV of this one.

We shall be wanting to know which of the six methods is (are) the most effective. We could proceed by comparing all possible combinations of groups, but that would be tedious since it would require, in this problem, fifteen such tests to check all the possibilities. What we need is some kind of "survey" test that will tell us whether there is any significant difference *anywhere* in an array of categories. If it tells us "no," there will be no point in searching further.

There are other reasons for using an overall test of significance that are more important in the long run than the saving of labor. First, any statistic based on *all* the evidence will be more stable (see Effect of N on Standard Errors, page 86) than one based on only part of it, as would be the case if only two of the six methods were compared. And second, there are so many comparisons that some will be "significant" by chance. If there were a hundred such comparisons, five would show significance at the .05 level and one at the .01 level, even if there were no real differences at all. So whenever we are dealing with several independent variables, we need an overall test of significance.

Such a test does exist. It is called the F test or F ratio. F is a ratio of two variances. If you will look back to Formula 4-2, page 28, you will find the standard deviation defined as

$$S = \sqrt{\frac{\Sigma x^2}{N}} \qquad (9\text{-}3)$$

The *variance* of a distribution, like its standard deviation, is a measure of variability. In the standard deviation formula, it is the part under the radical:

$$S^2 = \frac{\Sigma x^2}{N} \qquad (9\text{-}4)$$

It is the mean of the squared individual deviations, or, to put it another way, it is the square of the standard deviation.*

The *F* test is a ratio between two variance estimates. But before we proceed to analyze the logic of the *F* test, let's take just a moment to consider in very general terms what it is intended to accomplish. We have six groups. The mean of each group differs by some amount from that of every other group. The question is, "Are those *significant* differences?" (or more precisely, "Is there at least one significant difference among them?") We want to know whether the observed variability of the means is greater than could be expected by chance. Look at Figures 9-1 and 9-2. Which of these diagrams represents the more reliable (significant) differences among means?

Right. The smaller the variability of individual scores within each group, the more confident we can be that we really are dealing with different groups—or, more precisely, with samples representative of different populations. It is relatively difficult to imagine that the six groups in Figure 9-1 are random samples taken *from a single population.* That is the null hypothesis in analysis of variance—that all of the groups being compared are samples taken from the same population. The null hypothesis is intuitively more credible in Figure 9-2 than it is in Figure 9-1. But we must not rely on intuition; we need a quantitative test of the null hypothesis. For analysis of variance, that test is in a *ratio* known as *F.*

One important application of the *F* ratio is to an overall test of significance. As always in a significance test, the null

I   III   II   V   VI   IV

Figure 9-1 Groups I through VI with small variability within groups.

hypothesis states that all samples are random samples from a single population; and as always, we shall attempt to disprove that statement. Again, as in the z and t tests, it is a ratio that is being evaluated. But this time the critical ratio is not of a difference to its standard error; this time it is a ratio of two estimates of the population variance—one estimate from the means of the categories being studied, the other from individual scores *within* categories.

$$F = \frac{S_b^2}{S_w^1} \tag{9-5}$$

where S_b^2 = population variance extimated from observed variability among the means of the groups and S_w^2 = population variance estimated from observed variability within groups.*

You will remember that back in Chapter 7 (especially pages 80–82) we used obtained individual scores to make an estimate of the variability of means in a population. Well, it can work the other way, too; we can use obtained means to make an estimate concerning the variability of individual scores in a population. The numerator in the F ratio is just such an estimate[3]—the population variance estimated from the variability of *means* of the categories being studied. The denominator is the population variance estimated from the variability of *individual scores* within those categories.[4] So we have in this ratio two estimates of pop-

---

3. F is a statistic. You may wish to review pages 16–17 and 80–82 on the relation between a sample statistic and a population parameter.

4. The terms "category" and "group" will be used interchangeably, with the understanding that the referent in each case is a set of values derived ultimately from the behavior of individual subjects collected into categories for the purpose of the experiment.

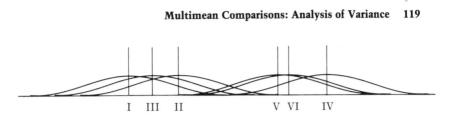

I   III   II   V   VI   IV

Figure 9-2   The same means, but with more variability within groups.

ulation variance. You could say that the numerator is an estimate of population variance with the effect of the independent variable included, while the denominator excludes that effect. If the two estimates are the same, there *is* no effect. The null hypothesis states that they *are* the same—that the ratio is 1.00.

If we are to demonstrate that some real difference does exist among the effects of our six teaching methods, we shall have to show that the ratio is not 1.00. Specifically, we shall have to show that the variance estimated from group means is greater than the variance estimated from individual scores within the groups. Understanding why that is so requires an analysis of the logic behind the *F* test.

Every estimate we make of population variance is an estimate of the variability of individual scores around a true mean. The closest thing we have to a true mean here is the *grand mean*. Every individual deviation from the grand mean can be thought of as two components: (1) the deviation of the individual score from its group mean and (2) the deviation of that group mean from the grand mean. Now, if we could somehow eliminate the means component, we should have an estimate of *what the population variance would have been* if there had been *no differences among the group means*. That estimate can be made by measuring the deviation of each individual score from its own group mean rather than from the grand mean and calculating the variance; the result is the denominator of the *F* ratio.

If you will compare Figures 9-3 and 9-4, you will see immediately that the total variability of the six groups is smaller when the variability of their means is eliminated. (You see only *one* mean in Figure 9-4, because all six are in the same place.) That is why the denominator of the *F* ratio is smaller than the numerator if there are any differences at all among the group means.

I   III   II   V VI   IV

Figure 9-3   Superimposed distributions of Groups I through VI.

Those group means do vary, almost always; the question is whether they vary enough to be significant in the sense that was developed in Chapter 8. To find out, we compute a ratio. To provide the denominator of that ratio, we discover how much *individuals* deviate from their own *group* means; whereas the numerator also takes into account the deviations of those *group* means from the *grand* mean.

The null hypothesis is that there is no variance among the-means: hence it predicts that a population variance estimate which *includes* that variance (the numerator of the F ratio) will be the same as one which *excludes* it (the denominator); if there are no differences among the means, F will equal 1.00. Ratios larger than 1.00 can be evaluated by tables that give the probability that any given F would occur entirely by chance. That, of course, is the *significance level* of the ratio, and it is interpreted in the same way as one which derives from a z or a t. We reject the null hypothesis if the probability of its occurrence by chance is sufficiently low; usually $p < .05$ will suffice.

After the F Test

When an F test turns out to be significant, we know (with some specified degree of confidence) that there is a real difference somewhere among our means. But we don't know *where* it is.

The most obvious approach to this problem is to perform a t test on each difference, beginning with the largest and continuing until one test fails to achieve significance. That is not acceptable, however, for some of the same reasons that prevented us from using t in the first place. Take another look at the third paragraph of the preceding section; the second reason cited there is the most cogent one for not using t after F. ("There are so many comparisons that some will be 'significant' by chance.") Some statisticians will approve the use of t in analysis of variance designs, either before

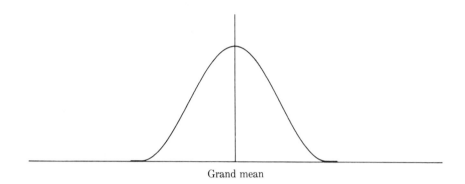

Grand mean

Figure 9-4 Superimposed distributions of Groups I through VI with differences between means eliminated.

or after analysis of variance, but only if the particular comparisons to be made are selected on a rational basis *before the data are collected*. It is the same requirement that was mentioned earlier in relation to the use of a one-tail test of significance (page 85); the reason for it is essentially the same, and the same controversy obtains.

Other tests have been devised for use in the post-*F* situation. All are attempts to disprove the null hypothesis, and all have been made more difficult to "pass" than *t* in order to compensate for the number of comparisons (and the concomitant increase in the probability that some will show significant differences by chance). Some of these tests are included in Appendix I under the title, Multimean Comparison Tests.

More Complex Designs

We have examined the logic of the *F* test using a "one-way" analysis of variance as an example. Others are possible: two-way, three-way, four-way, and so forth. My first impulse was to illustrate a two-way analysis by expanding our one-way example into a methods-by-teachers design; each method would have been used by five teachers, making 6 methods × 5 teachers = 30 treatment conditions. But that would have gotten us into problems inappropriate to a discussion of this kind. For example, the performance of a teacher using any one method probably would be affected by whatever experience he had had with the other methods; that would have taken us into the problem of counterbalancing the design. Also, the larger number of cells in such a matrix would have made your comprehension of *interaction* effects more difficult than necessary. So let's leave that one, with the passing comment that once the design problems have been solved and the treatments applied, data from such an experiment can be evaluated by analysis of variance techniques.

What we want now is the simplest design that could possibly be used to illustrate the basic principles of two-way analysis of variance. Such a design is called "2 × 2 factorial." design. Also, we'll select independent variables that do not require a counterbalanced design. As in the preceding illustration, there will be no computations.

The dependent variable in this experiment is "persistence in the face of failure"—specifically, the amount of time spent on an insoluble problem. The two independent variables are (1) high vs. low stress in the subjects and (2) success vs. failure experience immediately before the testing period. The design is shown in Table 9-5 as a 2 × 2 matrix in which there are four cells, each of which represents one treatment condition—(1) low stress with failure experience, (2) low stress with success experience, (3) high stress with failure experience, and (4) high stress with success experience. (Find the corresponding index numbers in Table 9-5). There are twenty-five subjects in each treatment condition; each subgroup is a random sample of a population of ten-year-old American males, and the question to be answered is whether they will *remain* random samples of a single population at the end of the experiment. The null hypothesis says they will.

Several hours before the experiment begins, all subjects receive a painful electric shock "accidentally" while playing with some laboratory equipment. (As will become apparent later, it is important that they know what it is like to be shocked.) Then, just a few minutes before the persistence task is introduced, they all take a short paper-and-pencil test. The test is also the same for all subjects; however, half of them (Group 2-4) are told that their performances are successful and the other half (Group 1-3) that their performances are not.

Immediately after that experience, they are all introduced to the persistence task. The problem is insoluble, but the subjects don't know that. They are all told that they should do the best

Table 9-5   A 2 × 2 Factorial Design with Index Numbers of Subgroups of Subjects as Cell Entries.

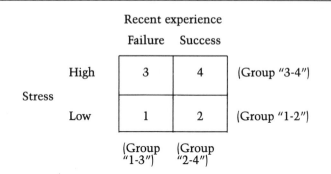

they can, but that they may leave at any time they wish. Then half of them (Group 3-4) are told that if they fail the test they will receive several shocks of the kind they had experienced earlier, while the other half (Group 1-2) are told nothing of the shocks; thus, we are manipulating stress as a second independent variable.

Now, analysis of variance offers three advantages over the z or t type of significance test: (1) it can compare the effects of more than two categories of an independent variable;* (2) it can compare the simultaneous but separate effects of two or more variables; and (3) it can assess the *interaction effects* of two or more variables. The first of those advantages was illustrated in the preceding section (One-Way Analysis of Variance). The second can be attained by merely repeating the F test for the second independent variable. In the present case, we would do one F test for the "stress" *main effect* (Group 1-2 vs. Group 3-4) and another for the "recent experience" main effect (Group 1-3 vs. Group 2-4). The most interesting of the three features of ANOVA, but also the most difficult to understand, is the last-named: the ability of this kind of analysis to identify interaction effects.*

The best way to explain interaction is to cite an example, so let's return to our experiment. Table 9-6 is like Table 9-5, except that the index number of each subgroup has been moved to the upper left-hand corner of that subgroup's cell, and the number in the middle of the cell is the subgroup's mean score (time spent before quitting). The data in the table show what look like two substantial main effects. (Whether they are *significant* depends,

Table 9-6 Group Means (Minutes at Task) in Two-Way Analysis. Outcome 1.

		Recent experience	
		Failure	Success
Stress	High	[3] 20	[4] 30
	Low	[1] 10	[2] 20

of course, on the amount of variance *within* the two groups that are being compared in each case; let us assume a sufficiently small amount.) Group 3-4 (50 minutes) is more persistent than Group 1-2 (30 minutes), and Group 2-4 (50 minutes) is more persistent than Group 1-3 (30 minutes). High stress produces more persistence than low, and success more than failure. But there is no interaction.

By contrast, imagine the outcome of the experiment as that depicted in Table 9-7. There is *one* main effect (success vs. failure), and *there is interaction* because increasing stress (from low to high) has a different effect on subjects in the "failure condition" (down 10) than it does on those in the "success condition" (up 10). Similarly, success (as opposed to failure) has a different effect in the "high-stress condition" (up 20) than it does in the "low-stress condition" (no change). The interaction effect may be seen more clearly in the graphical representation of the two outcomes shown in Figures 9-5 and 9-6.

Table 9-7 Group Means (Minutes at Task) in Two-Way Factor Analysis. Outcome 2.

		Recent experience	
		Failure	Success
Stress	High	[3] 10	[4] 30
	Low	[1] 20	[2] 20

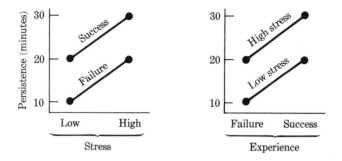

Figure 9-5   Graphic representation of Outcome 1, showing two main effects and no interaction.

Very different patterns, are they not? In Outcome 1, there is a very simple main effect of stress shown in the left diagram and of success on the right. Outcome 2, however, shows that the effect of high stress—as opposed to low—is to *increase* persistence in subjects who have recently experienced success and to *decrease* it in those recently subjected to failure. It shows also that the effect of an experience of success—as opposed to failure—is to increase persistence dramatically in the high-stress group but to make no change at all in the low-stress group.*

We might be tempted to speculate about possible explanations for those results. But in the first place, this is a treatise on statistics, not psychology, and in the second place, these data were

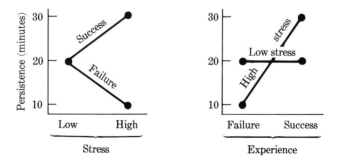

Figure 9-6   Graphic representation of Outcome 2, showing interaction and one main effect.

not derived from any real experiment. The main point is that interactions do occur, and that when they do, they can be detected by analysis of variance.

Summary

Chapter 8 showed how a difference between two groups of subjects can be evaluated in terms of the probability of its occurrence by chance (sampling error). Chapter 9 has extended that technique to situations in which the data are in the form of *frequencies* and to those in which *several* groups vary on a given factor (independent variable)—and even to some in which several factors must be evaluated simultaneously.

The logic of the *F ratio* is revealed in the course of its application to a *one-way analysis of variance* in a six-category experimental design. An example is given also of a more complex design to which analysis of variance might be applied—a two-way analysis in a 2 × 2 *factorial design*. A distinction is drawn between *main effects* and *interaction effects*. Because *F* functions initially as a kind of "survey" test, the follow-up problem—the identification of the precise source of these effects when *F* has proven significant—is also discussed.

Appendix I attempts to convey an appreciation of the breadth of application of significance tests and, simultaneously, to provide you with a ready reference list, by naming a large number of such tests and identifying their functions.

SAMPLE APPLICATIONS FOR CHAPTER 9

EDUCATION

1. In an attempt to decrease the number of high school dropouts, a school system develops a vocational training program for students who are at high risk for becoming dropouts. During the first several years of the program, the school staff wants to determine whether it is effective in reducing the number of students who drop out, and you are asked to help. You select at random the fifty students that the program can accommodate from all students who apply and are "at risk." Then you keep track of the number of students who eventually graduate. What statistic will help you decide whether the special program is effective?

2. You are a school psychologist. You have developed a program to help students systematically think about and plan solutions to social problems that they may be faced with in school, in dealing with friends, and in job situations. To try it out, you train half of the counselors in the school system in the use of the program. Teachers first identify students who are having difficulty solving social problems. Then, one-third of those students is assigned randomly to trained counselors and another third to untrained counselors. The remaining third is provided no counseling at all. At the end of the semester, all are tested to determine their ability to solve a number of social problems. How will you interpret the resulting data?

3. You are an educational psychologist. You and a group of your colleagues have developed four different educational programs to enhance the abstract reasoning ability of sixth- and seventh-grade students. With the cooperation of a large urban school district, 200 sixth- and seventh-grade classrooms are assigned to the four treatment programs, fifty classrooms per program. All

students are tested on abstract reasoning at the beginning and end of the year. How could the resulting data be analyzed and interpreted?

POLITICAL SCIENCE

1. You are interpreting the results of an opinion poll. In a random sample of 100 individuals you find that of the 40 Republican respondents a total of 10 supported the SALT II treaty, and that of the 60 Democratic respondents 30 supported the treaty. How can you calculate the probability that this association is due to chance?

2. Once again you are studying the incidence of military coups in Latin America. This time you want to know whether there are significant differences in the frequency with which coups occur, and you have reason to believe that the type of legitimacy upon which the regime rests (i.e., traditional, legal-rational, charismatic) is the most important factor in accounting for the incidence of coups. In this situation you have an independent variable (type of regime legitimacy) and a dependent variable (coup frequency). How can you demonstrate that types of regimes probably do (or do not) differ with respect to coups?

3. One of the most frequently studied questions in the field of international relations pertains to the relationship between domestic politics and foreign policy behavior. You are in that field and you want to test the idea that a nation's form of government (democratic, authoritarian, or totalitarian) and type of leadership (unitary, collective, or fragmented) affects the number of aggressive actions it initiates (the dependent variable). Since there are two independent variables here instead of one, what kind of analysis is appropriate?

PSYCHOLOGY

1. You are the director of a clinic dealing solely with phobic behaviors (unusual fears) of childhood. Over a two-month period, 100 children are referred to your clinic for the treatment of agoraphobia (fear of large open spaces). After hypnosis therapy, you find that 55 of the 100 children are able to walk in a

large open field and report little or no fear. These are fairly impressive results, but are they statistically significant? How can you tell?

2. You are a child clinical psychologist interested in which of three approaches is most effective in controlling the pain of children in a hospital burn unit. You assign, at random, equal numbers of children to (a) a mental distraction condition, where the participants attempt to keep their minds off the pain by doing mental arithmetic; (b) a self-reward condition where participants give themselves positive self-statements (e.g., "I'm a brave person") for enduring pain; and (c) imagination exercises where the participants are taught to imagine situations in which they experience pleasure and satisfaction. How might you go about analyzing the results?

3. You are a psychotherapist at a child guidance center. You are interested in the possible interaction between personality factors and the effectiveness of psychotherapy. You assign each of twenty *introverted* (shy; socially withdrawn) children to either *individual* or *group* therapy. Next, you assign twenty *extraverted* (socially outgoing) children to the same conditions. Your prediction is that the introverted children, because of their social fears and shyness, will benefit most from individual therapy and that the extraverts will work best in a group setting. Presuming that you have a generally accepted criterion of "effectiveness of therapy," how might you analyze the data from this study?

SOCIAL WORK

1. You are a social worker in a small rural community. You conclude that the existing manner of dealing with juvenile offenders through the courts is of limited efficiency and effectiveness. The magistrate is present only one and one-half days a week, and the court calendar is always too full to handle the number of youth being petitioned into court.

You develop an alternative to this process by convincing local citizens and the judge to establish a Juvenile Review Board. The Board, composed of community people, will decide cases and use measures such as community work service, restitution, and active parental involvement instead of the jail time and probation that have been the remedies usually prescribed by the courts. Part of the agreement in establishing the alternative program is an evaluation which will assess its effectiveness, at the end of one year, in comparison with the courts' traditional procedures. Recidivism is selected as the major indicator of success and youth will be assigned to the alternative program or to the courts on a random basis, with violent offenders excluded from both groups.

When the one-year period is over and all the data are in, how do you evaluate the results?

2. In your Child Guidance Clinic there is disagreement concerning how to treat children who display hyperactive behavior. One social worker with a psychoanalytical background favors play therapy with two sessions a week for a minimum of one and one-half years. The agency psychologist favors a behavioral model and the use of operant conditioning. A consulting psychiatrist views hyperactivity as the result of immature cortical development, and she recommends the drug methylphenidate to stimulate the cortex.

Since hyperactivity is so frequent and so troublesome, you recognize it as a major problem. You decide to conduct a study to see which is indeed the most effective of the three treatment methods; you also decide to add a fourth group of children who will receive no treatment. (The clinic has a waiting list.) Children will be assigned randomly to those four groups. The social worker, the psychologist, and the psychiatrist all agree on a single test score as the criterion of effectiveness. Once you have obtained all your data, how do you evaluate it?

3. As a social worker in a foster care agency, you are interested in how the age of children and youth affects their adjustment to either a foster family home or a staffed community group home.

The independent variables are *age* and *type of foster care* arrangements. Age is defined by two categories: children (5 to 12 years) and youth (13 to 18 years). You obtain a standardized instrument to measure the dependent variable, *adjustment* to foster care.

Twenty children are assigned on a random basis to either a foster family home (10 children) or a group home (10 children), and twenty youth are assigned on a random basis to either a family home (10 children) or a group home (10 children).

SOCIOLOGY

1. You are still consultant to the family-planning group described in the sociology exercise for Chapter 8. This time you are asked about the antecedents of different attitudes toward abortion. You cannot answer such a broad question all at once, but one of your many hypotheses is that a person's belief about when human life begins influences his attitude toward abortion. Some questionnaire data are available from another study, and you are able to find two questions pertinent to your hypotheses: (1) "Do you believe that human life begins before or after the ninetieth day following conception?" and (2) "Do you approve of abortion on demand?" The respondents' answers to these questions are your data. What do you do with them?

2. That family-planning agency is intrigued by your answer to its earlier question (Sociology, Chapter 8) about the effect of religious belief on family size. The agency wants to broaden the investigation by comparing the mean sizes of Catholic, Mormon, and Mennonite families with each other and with that of the general population. You gather the data and find that there are differences among the means. How can you tell whether these differences are significant?

3. You are asked to identify some of the critical variables that inhibit or enhance communication of sex information to children by their parents. "Amount of sex information communicated" is therefore the dependent variable in your study. You construct an instrument which yields a sex information score and administer it to parents of ten-year-old children. (The score represents the amount of information that parents believe they would give to their ten-year-old if the child were to ask, "Where do babies come from?") You hypothesize that the amount of information contained in the parents' answers to this question is (1) inhibited by church involvement and by lack of education and (2) enhanced by little or no church involvement and high education. You define *Church Attendance* as attendance four times a year or more, *Nonattendance* as attendance of three times a year or less. One or more years of college defines the *High Education* group, and less than a year of college identifies the *Low Education* group. How will you analyze the data that emerge from this study?

10 | SUMMARY

The central objective of this book has been to convey an understanding of basic concepts and their interrelations—the "big ideas," so to speak, in statistical thinking. In the last two chapters, for example, the "big idea" was the logic of testing for the significance of differences, especially in experimental settings; in Chapter 5, it concerned the description of relationships among variables in settings other than experimental.

The first four chapters established a necessary foundation for everything that was to follow; but in Chapters 2, 3, and 4 it became apparent that "big ideas"—for example, the idea of "distribution," the importance of specifying the kind of average being used, the concept of "variability"—are involved even in the description of a single sample. Chapter 6 stressed the importance, when interpreting any measurement of a given subject's behavior, of comparing it to similar measurements that have been made of some identifiable reference group. The central concern of Chapter 7 was the relation between a sample and its population.

Those are all important ideas. They won't enable you to *do* social science research, but they will give you access to research done by others. That is clearly a high enough payoff to have justified all the effort you have expended in mastering them; but there is an additional reward which, though incidental to the stated objectives of the book, could well prove most important of all in the end. While learning to think better *statistically*, you may have learned to *think* better!

You may not have occasion to "think statistically" every day, every week, or even every month; but whenever the occasion does arise, you will be ready for it. After an especially long hiatus, you may not be able to deal immediately with every concept that you have mastered while studying the book; but you will find that a quick reference to its discussion of a particular target concept will revive your understanding not only of that concept, but of many related ones as well.

General agreement is so far lacking on a single set of symbols to refer to the various concepts that have been developed here. For that reason, I have provided a conversion table (Appendix II) that will allow you to interpret any of the commonly used statistical symbols that you may encounter in your reading.

Good luck! And remember what Sir Francis Galton used to say: "Whenever you can, count."

A P P E N D I X   I | TESTS OF SIGNIFICANCE

Virtually any kind of a difference can be tested for statistical significance. The only requirement is that the data be expressed numerically. Much of the research—especially the experimental research—that concerns social scientists and professionals is evaluated in those terms.

This appendix lists a large number of significance tests. Not all of them are labeled as tests of "differences'; they are such, nonetheless, because they are all testing the hypothesis of no (null) deviation from some specified point of reference, and the question in each case is whether the obtained deviation differs from zero.

Do not allow yourself to be frightened by the following formulas. Look at them first with the idea of discovering how broad is the application of significance testing; that is, indeed, the major reason for including them here. Later, you may wish to use this list as a handy reference, as you encounter various significance tests in your reading. Some of the tests are described partly or wholly in words, others entirely by their formulas. If the formulas—or the verbal descriptions, for that matter—don't make sense to you, don't worry about it. They will become more meaningful within the context of a research report. On the other hand, if you enjoy a challenge, you might want to try reading and conceptualizing the formulas (that is, comprehending the relationships expressed in them). I have provided some assistance in the form of labels for unfamiliar terms.

In all formulas, the effects of sample sizes on variance estimates are represented by degrees of freedom rather than N.

The *t* Tests

1. Correlation (r)

$$t = r\sqrt{\frac{N - 2}{1 - r^2}}$$

2. Partial correlation (of two variables, with the influence of one or more others partialed out) $(r_{12\cdot3} =$ Correlation of variables 1 and 2, with variable 3 partialed out)

$$t = r_{12\cdot3}\sqrt{\frac{N - 3}{1 - r^2{}_{12\cdot3}}}$$

3. Rank-order correlation $(\rho, \text{ or rho})$

$$t = \rho\sqrt{\frac{N - 2}{1 - \rho^2}}$$

4. Difference between correlated correlations $(r_{12} \text{ vs. } r_{13})$

$$t = \frac{(r_{12} - r_{13})\sqrt{(N - 3)(1 + r_{23})}}{\sqrt{2(1 - r_{12}{}^2 - r_{13}{}^2 - r_{23}{}^2 + 2r_{12}r_{13}r_{23})}}$$

5. Difference between regression coefficients $(\beta, \text{ or } beta)$ $(S_{d_\beta} = \text{Standard}$ deviation of differences between betas)

$$t = \frac{\beta_1 - \beta_2}{S_{d_\beta}}$$

6. Difference between obtained and hypothesized $(\overline{X}_h)$ means

$$t = \frac{\overline{X} - \overline{X}_h}{SE_{\overline{X}}}$$

7. Difference between independent means

$$t = \frac{\overline{X}_1 - \overline{X}_2}{\sqrt{\dfrac{S_1{}^2}{N_1} + \dfrac{S_2{}^2}{N_2}}}$$

8. Difference between correlated means

$$t = \frac{\overline{X}_1 - \overline{X}_2}{SE_{\overline{X}_1 - \overline{X}_2}}$$

9. Difference between independent variances

$$t = \sqrt{\frac{S_1^2}{S_2^2}}$$

10. Difference between correlated variances

$$t = \frac{(S_1^2 - S_2^2)\sqrt{N - 2}}{\sqrt{4S_1^2 S_2^2 (1 - r_{12}^2)}}$$

The *F* Tests

1. Difference between variances of two distributions

$$F = \frac{S_1^2}{S_2^2}$$

2. Analysis of variance

    (a) Treatments (entire matrix) (S_T^2 = Treatment variance, entire matrix; S_e^2 = Error variance, entire matrix)

$$F = \frac{S_T^2}{S_e^2}$$

    (b) Main effects (by factors) (S_A^2 = Factor *A* variance between *A* groups averaged over *B* levels; S_e^2 = Error variance within *A* groups)

$$F = \frac{S_A^2}{S_e^2}$$

    (c) Interaction (entire matrix) (S_I^2 = Interaction variance, various combinations; S_e^2 = Error variance, entire matrix)

$$F = \frac{S_I^2}{S_e^2}$$

    (d) Trends of means

        (1) Linearity (S_L^2 = Linear component; S_e^2 = Error variance, entire matrix)

$$F = \frac{S_L^2}{S_e^2}$$

        (2) Curvature (S_Q^2 = Quadratic component; S_e^2 = Error variance, entire matrix)

$$F = \frac{S_Q^2}{S_e^2}$$

3. Eta (η) correlation coefficient for curved regression line (G = Number of groups)

$$F = \frac{\eta^2/G - 1}{(1 - \eta^2)/(N - G)}$$

4. Multiple correlation (R) of one variable with a weighted composite of two or more others (m = Number of constants in the regression)

$$F = \frac{R^2/m}{(1 - R^2)/(N - m - 1)}$$

5. Beta (β) weight in a multiple regression equation ($R_{1 \cdot 23}^2$ = Multiple correlation of variable 1 with 2 and 3)

$$F = \frac{R_{1 \cdot 23}^2 - R_{13}{}^2}{(1 - R_{1 \cdot 23}^2)/(N - 2 - 1)}$$

6. Coefficient of concordance (W) (Correlation coefficient indicating agreement among judges) (m = Number of judges)

$$F = \frac{W(m - 1)}{1 - W}$$

Non-Parametric or
Distribution-Free Tests[1]

1. Chi square (χ^2).

2. Friedman analysis of variance by ranks (two-way).

3. Coefficient of concordance (agreement of judgments).

4. Kruskal-Wallis analysis of variance (one-way). Called H test.

5. Median test. χ^2 used after division of groups at median.

6. Sign test. Positive vs. negative changes when experimental sub-jects are compared to matched control subjects. (Null hypoth-esis = same number of positive as negative changes.)

7. Wilcoxon test. Positive vs. negative changes, as above.

8. Mann-Whitney U test. Alternative to the t test when popula-tions lack normality of distribution and/or homogeneity of var-iance.

9. Kilmogorov-Smirnov test. Goodness of fit of a theoretical to an observed frequency distribution. Alternative to chi square when N is small.

---

1. The tests in this section are used (1) when the data come from discrete (as opposed to continuous) categories or (2) when quick estimates are needed. When these tests are used with continuous data there is a loss of information similar to that which occurs when the Spearman rho is used in place of the Pearson r as a coefficient of correlation. The tests are called "distribution-free" because they do not involve any assumptions concerning the distribution of the refer-ence population, whereas, with the single exception of chi square, all of the significance tests discussed in this book were developed originally on just such assumptions. Since those assumptions now appear to be far less important than originally thought, the recent trend toward increasing use of distribution-free tests is currently being reversed.

Multimean Comparison Tests

These tests are used after a significant main-effect F in analysis of variance.[2]

1. Duncan's multiple range test. Means are arranged in a hierarchy, and ordered comparisons made (largest minus smallest mean, largest minus second smallest, and so on). Significance is based on the number of ordered steps between means. It is possible to test any and all means, taken two at a time. Each comparison must be planned in advance.

2. Newman-Keuls method. Ranked means or sums of experimental interest are tested for significance, based on the number of ordered steps involved in the comparison. Planned in advance.

3. Studentized Range statistic. Test for significance of largest minus smallest means only. Planned in advance.

4. Dunnett test. Comparison of each mean with a control or standard mean.

5. Scheffé test for any and all mean or sum comparisons. Does *not* need to be planned in advance.

6. Orthogonal test for comparison between independent sets of means. Planned in advance.

7. Tukey tests. Tukey has devised techniques for locating the source of significance in an array that has produced a significant F. For details, see J. W. Tukey, "Comparing individual means in the analysis of variance," *Biometrics*, 1949, vol. 5, pp. 99–114.

---

2. Each of these tests is designed to ascertain which differences among groups are significant.

A P P E N D I X   II

LIST OF SYMBOLS

The table on the following page shows the symbols used in this book as well as alternative notations you may encounter in other statistics textbooks. Though the symbols used in this book may be the most common, there is no standard usage either accepted by authors in the field or approved by the American Statistical Association. [Personal correspondence with John W. Lehman, Executive Director, American Statistical Association, 26 June 1972.]

Sample statistic		Population parameter used in this book	Name or description
Used in this book	Some alternative notations		
X			Raw score
x	d		Deviation score
Σ			Summation
N	n		Number of observations
$\overline{X}$	M, x	μ	Mean
S	s, SD or SD, σ	σ [1]	Standard deviation
S^2	s^2, V, σ^2	σ^2 [1]	Variance
$SE_{\overline{x}}$	$\sigma_{\overline{x}}, \sigma_M, S_{\overline{x}}$, SE or SE	$\sigma_{\overline{x}}$ [1]	Standard error of the mean
$SE_{\overline{x}_1 - \overline{x}_2}$	$SE_x, \sigma_{(\overline{x}_1 - \overline{x}_2)}, \sigma_{d_M}$	$\sigma_{\overline{x}_1 - \overline{x}_2}$ [1]	Standard error of a difference
ρ		[2]	Rank-difference correlation
r		[3]	Product-moment correlation
z t			Ratio of a quantity to its standard deviation or standard error
F			Ratio of one variance to another
χ^2			Chi square
p	P		Probability of wrongfully rejecting the null hypothesis. Level of significance.

1. Only verbal reference is made to this concept in the text, but there is a fairly well-established convention that this parameter symbol should be used as a companion to the *statistic* symbol in the first column.

2. This system of notation calls for a transformation to Greek; since the statistic ρ is already Greek, that can't be done here.

3. Here the system calls for a ρ; but that would be confusing, because ρ already has another meaning.

NOTES

The following supplemental notes are indicated in the text by asterisks.

Chapter 3

Page 14
A figure like this—essentially a vertical bar graph—is called a *histogram*. A histogram is one way of representing a frequency distribution.

Page 16
The distribution of means would be normal even if the samples were not.

Page 18
If you should ever have to do the computations yourself, you would find that the median is often *within* a class interval rather than precisely between two intervals as in the above illustrations. In that situation, it is necessary to interpolate; any methods text will quickly tell you how.

Page 20
A complication arises when the distribution includes the scores of two "types" of individuals. That can result in two or more modes (in which the distributions are called *bi*modal, *tri*-modal, etc.). Multimodality should never be interpreted unless the N is rather large, however; for in a small distribution, chance

deviations can easily produce a "humpy histogram." ("Histogram" is defined above in the first note on this chapter.)

Chapter 4

Page 28

Many statisticians use the quantity $N - 1$ in place of N when computing sample variance. $N - 1$ gives the number of *degrees of freedom* in the calculation of the statistic. (In other applications, there might be $N - 2$, $N - 3$ or $N - n$ degrees of freedom.) The calculation of the standard deviation requires the previous calculation of a mean, and one degree of freedom is "used up" thereby.

The term *degrees of freedom* refers to the number of scores that are free to vary. For example, imagine a very simple situation in which the individual scores that make up a distribution are 3, 4, 5, 6, and 7. If you are asked to tell what the first score is without having seen it, the best you can do is a wild guess, because the first score could be *any* number. If you are told the first one ("3") and asked to give the second, you probably won't guess some number in the millions, but logically there is no reason why it should *not* be in the millions or even higher; it, too, can be any number. The same is true of the third and fourth scores; each of them has complete "freedom" to vary. But if you know those first four scores (3, 4, 5, 6) *and you know the mean* of the distribution (5), then the last score can only be 7. If instead of the mean and the 3, 4, 5, and 6, you were given the mean and, say, 3, 5, 6, and 7, the missing score could only be 4. In any case, if the mean is known, the missing score is *determined* by your knowledge of the other four, and the *degrees of freedom* of all the scores taken together has been reduced by 1; hence the term "$N - 1$" in the formula for variance.

Our calculation of the mean was done directly from independent samples of the population (see pages 13–17 and 86–87). The number of degrees of freedom was therefore equal to N. But when we work with deviation scores, as we do in the average deviation and the standard deviation, degrees of freedom is less than N because the deviations are from a mean, and the calculation of a mean uses up one degree of freedom, as indicated above.

Since all our statistics are based upon the assumption of independent ("free") observations, any statistic that does not make a correction for loss of independence, or "freedom," is a biased estimate of its population parameter.

In this book, that bias will be overlooked, and N will be used everywhere as degrees of freedom. There are four reasons for so doing. The first is that the main text carefully develops the concepts "mean of the raw scores"

$$\frac{\Sigma X}{N}$$

"average deviation"

$$\frac{\Sigma |x|}{N}$$

"standard deviation"

$$\sqrt{\frac{\Sigma x^2}{N}}$$

and finally (on page 117), "variance"

$$\frac{\Sigma x^2}{N}$$

as a sequence of constructs characterized by a common structure; to have changed N to $N - 1$ in the second formula would have broken the sequence. The second reason for omitting that refinement is that it does not contribute to an understanding of any of the above mentioned statistics as measures of variability. A third reason for leaving it out is that even writers of more advanced texts often use

$$\sqrt{\frac{\Sigma x^2}{N}}$$

as their measure of variability in the sample, with or without apologies to their colleagues. And finally, if degrees of freedom $= N - 1$, as is the case of every statistic discussed in the main body of this text, the difference between "N" and "degrees of free-dom" is negligible, providing only that the sample is not extremely small. With a sample of only two scores, for example, $N - 1$ would be only *half* as big as N; with a sample of 100, on the other hand, $N - 1$ would be only 1 percent smaller than N.

Page 31

The semi-interquartile range is a much more precise statistic than the total range, but no adjustment is made *there* for the ½ unit cut off at either end of the Q_1 from Q_3 interval in the process of subtracting Q_1 from Q_3. In the case of the total range, my own view is that the only time that such an adjustment is really nec-essary is when the question to be answered is specifically, "What is the smallest interval within which all cases will fit?" If you were to refer to "elementary school grades one through six," for example, it would be misleading to speak of a range of five grades (largest = 6 minus smallest = 1).

Chapter 5

Page 38

Actually, a Pearson r (pages 41–50) coefficient on these data would be slightly less than 1.00. The relationship is perfect, but r is accurate only in rectilinear (straight-line) regressions, and this one is really curved. (For every height increase of a uniform amount, the corresponding weight increase is systematically greater as you move from left to right in Figure 5-1.) Another coef-ficient can be used in such cases; look for "eta (η) coefficient" or "correlation ratio" in any text on statistical methods.

Page 39

If you *should* ever need a quick estimate of correlation, rho is the one to use. Consult any standard statistical methods text about what to do with tied ranks, and within a few minutes you'll be ready to use formula 5-1.

Page 41

Those assumptions are : (1) that each of the two distributions is unimodal and symmetrical, and (2) that the line best repre-senting the relation between them is straight rather than curved.

(See the diagrams in the Scatterplot section of this chapter, pages 46–49, for examples of such "rectilinear" relationships. Also see the preceding note concerning curved ones.) Actually, recent computer simulations (Havlicek, L. L. and L. Peterson, "Effect of the Violation of Assumptions Upon Significance Levels of the Pearson *r*," *Psychological Bulletin*, 1977, vol. 84, no. 2, pp. 373–377) have shown that violation of traditional assumptions does not affect *r* very much. It is therefore said to be a *robust* statistic.

Page 50
A sample must be taken at random or by some other approved method; but even when the "simple" random method (see Footnote 1, page 4, for definition) is not used, randomness characterizes an important part of the total selection procedure.

Chapter 6

Page 66
Originally, a *T* score was any standard score other than a *z* score, but usage has gradually changed its meaning to "a standard score in a distribution that has a mean of 50 and a standard deviation of 10."

Page 75, first note
Usually median.

Page 75, second note
And his IQ is 100. Divide mental age by chronological age,

$$\frac{\text{MA}}{\text{CA}} = \frac{6}{6} = 1.00$$

Then drop the decimal point.

Page 75, third note
And his IQ is 133 (8/6 = 1.33).

Page 75, fourth note
See "Deviation IQs" in Figure 6-3. The IQ defined in the second note (above) for page 75 was for many years *the* IQ. Now, however, it is known as the "ratio IQ" to distinguish it from the

newer "deviation IQ." (See "Deviation IQs" scale in Figure 6-3). A deviation IQ is a standard score. In a distribution of general mental ability test scores, the obtained mean is converted to 100, because the mean ratio IQ is 100. Then the standard deviation of the raw scores is changed to match that of ratio IQs on the same test. The result is a set of standard scores almost the same as those which would have been obtained by the old method. The new method is easier to use, and it has other advantages the explanation of which would require a more extensive digression into psychometrics than would be appropriate here.

Chapter 8

Page 101, first note
This is the formula used when means are uncorrelated. If students were *matched* on some variable—for example, intelligence—related to their level of performance under both control and experimental conditions, a correlational factor would have to be introduced. But that refinement is beyond the scope of this book, and the basic idea of

$$SE_{\bar{X}_e - \bar{X}_c}$$

is better conveyed by Formula 8-1 as it stands.

Page 101, second note
To find how many standard errors an obtained difference is from zero, it is necessary to find the size of the standard error. In this instance, there were two such computations:

First, there was the standard error of the difference for Figures 8-3 and 8-4, assuming that the standard errors of the means of the two distributions are as indicated in Figures 8-1 and 8-2:

$$SE_{\bar{X}_c} = 2; SE_{\bar{X}_e} = 2$$
$$SE_{\bar{X}_e - \bar{X}_c} = \sqrt{SE_{\bar{X}_c}^2 + SE_{\bar{X}_e}^2}$$
$$SE_{\bar{X}_e - \bar{X}_c} = \sqrt{2^2 + 2^2} = \sqrt{4 + 4} = \sqrt{8} = 2.83$$
$$SE_{\bar{X}_e - \bar{X}_c} = 2.83$$

Second, there was the standard error of the difference for Figures 8-7 and 8-8, assuming that the standard errors of the means of the two distributions are as indicated in Figures 8-5 and 8-6:

$$SE_{\bar{X}_c} = 5; SE_{\bar{X}_e} = 5$$
$$SE_{\bar{X}_e - \bar{X}_c} = \sqrt{SE_{\bar{X}_c}^2 + SE_{\bar{X}_e}^2}$$
$$SE_{\bar{X}_e - \bar{X}_c} = \sqrt{5^2 + 5^2} = \sqrt{25 + 25} = \sqrt{50} = 7.07$$
$$SE_{\bar{X}_e - \bar{X}_c} = 7.07$$

Page 103

The "significance level" states the probability that we are making an error when we reject the null hypothesis. That kind of error is known as "Type I." The probability of making a Type I error is sometimes called "alpha error" or simply "alpha" (α). By stating before the experiment that we will accept only a very low probability (that is, a low probability of rejecting a null hypothesis that is in fact true), we can reduce our Type I errors to a level approaching zero. Unfortunately, however, the *lower* we set *that* level, the *higher* is the probability of *accepting* a null hypothesis that is in fact *false.* The latter kind of error is called "Type II." The following table summarizes those relationships. The ability of a significance test to make the decision represented by the lower left cell of the table—that is, to reject a false null hypothesis—is called its "power."

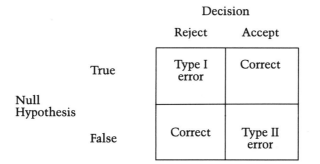

		Decision	
		Reject	Accept
Null Hypothesis	True	Type I error	Correct
	False	Correct	Type II error

Page 104

Many authorities will not accept a one-tail test under any circumstances. Their argument is that we cannot have a good reason for expecting one group to be superior to the other—or rather, that if we did we wouldn't have to make the statistical test. To put it another way, the reason that we conduct the test at all is precisely that we *don't* know what the outcome will be. There is controversy in every field.

Chapter 9

Page 111

In our examples, the "no difference," or "no deviation," hypothesis specifies no deviation from a difference of zero between (or among) treatment groups. By extension, the "null hypothesis" has become *any hypothesis concerning the location of a parameter*—a hypothesis of no difference *from that hypothetical location*, whether or not the location is zero.

A good illustration of a non-zero null hypothesis comes from "quality control" in industry. If the product is supposed to contain x amount of some chemical, for example, then each random sample is tested against a null hypothesis of x, and any "significant" ($p < .05$?) deviation from x requires remedial action.

Page 112

Technically, of course, it is the null hypothesis that is being tested rather than the obtained difference. But when such a test reveals that the null hypothesis is probably not true, most researchers allow themselves to speak of the *difference* as significant.

Page 115, first note

If instead of a simple "yes" or "no," we were to use a ten-point scale extending from "strongly approve" to "strongly disapprove," we could use this formula without alteration; as it is, accuracy demands that we correct for the crudity of our two-point scale. But my purpose in this book is to reveal the basic structures of statistics. I have therefore declined to present to you a formula that disguises the structure on which chi square is based. If you are interested in computing a chi square, you will need a "correction for continuity" (really a correction for the *discontinuity* imposed by a crude scale of measurement); you will find one in any good text on statistical methods.

Page 115, second note
The computation may be of concern to some readers. Here it is:

$$\chi^2 = \Sigma \frac{(f_o - f_e)^2}{f_e}$$

$$\chi^2 = \frac{(50 - 60)^2}{60} + \frac{(50 - 40)^2}{40} + \frac{(70 - 60)^2}{60} + \frac{(30 - 40)^2}{40}$$

$$\chi^2 = \frac{-10^2}{60} + \frac{10^2}{40} + \frac{10^2}{60} + \frac{-10^2}{40}$$

$$\chi^2 = \frac{100}{60} + \frac{100}{40} + \frac{100}{40} + \frac{100}{40}$$

$$\chi^2 = 1.67 + 2.5 + 1.67 + 2.5$$

$$\chi^2 = 8.34$$

Table III of R. A. Fisher, *Statistical Methods for Research Workers* (Edinburgh and London, Oliver and Boyd, 1932), gives 6.635 as the χ^2 required for significance at the .01 level; so our 8.34 is significant at an even higher level—that is, at *less* than .01.

Page 117
An alternative to the S^2 notation is V. But there is a more important alternative to the formula itself: if you have read the note (above) for page 28, you know that the quantity used as the denominator is often $N - 1$ rather than N.

Page 118
This estimate is sometimes referred to as "residual," or even "error," variance (S_e^2). The latter term may be somewhat confusing, because most of the variance probably stems from factors that in different circumstances might prove to be effective independent variables. They are regarded as "error" in the context of a given investigation because they are random with respect to the independent variables in *that* experiment. They are the so-called "con-

trolled" variables. There are two general strategies for "controlling" a variable in an experiment: (1) to hold it constant over all conditions of the independent variable(s) and (2) to randomize its variance. Both strategies are designed to achieve the same objective: prevention of any systematic relationships between the "controlled" variable and any independent variable. There usually are several controlled variables in an experiment.

Page 123, first note

When only two groups are being compared, the analysis of variance is really a *t* test; or to put it another way, *t* is a special case of analysis of variance that can be used only when there are just two categories of the variable being examined. In such a case, $t = \sqrt{F}$; with large samples, of course, the same is true of *z*—that is, $z = \sqrt{F}$.

Page 123, second note

The *F* ratio for interaction is

$$F = \frac{S_I^2}{S_E^2}$$

where S_I^2 = interaction variance and S_E^2 = error (within-group) variance, both derived from the entire matrix.

Page 125

In order to achieve maximal simplicity—and hence clarity—in this example, I have assumed the linearity of each of the functions depicted in Figures 9-5 and 9-6. A more refined experimental design might have defined those functions more precisely by including *several degrees* of stress (instead of just two) and *several degrees* of success. If that were done in the case of stress, for example, it might turn out that *medium* amounts of stress produce more persistence than either extreme amount. An investigation of the effects of a single variable through many changes in its value is sometimes referred to as a *parametric* study.

SOLUTIONS TO SAMPLE APPLICATIONS

Chapter 3

Education

A good choice. The mean of the students' reading level gives you the most stable estimate of the average reading level of the 2000 students.

Possible misinterpretations. Misinterpretations include finding a single central tendency when there are really two. It is possible that students enrolled in the semi-skilled trades (such as food service, construction, sewing, and horticulture) and students in the skilled trades (such as computer programming, electronics, radio and television) have very different reading abilities. If so, the mean reading level of the total sample falls between the two subgroups. It therefore does not represent either of them, and materials purchased on the basis of the mean would be at too high a level for one group of students and too low a level for the other.

Political Science

A good choice. The mean can give you an estimate of the "balance point" of the distribution of European military spending. Of all the measures of central tendency, the mean is the one that would vary least if repeated random samples were taken.

Possible misinterpretations. Despite the general virtues of the mean as a measure of central tendency, it may provide an inaccurate picture of a particular situation. The mean is highly sensitive to any single deviant case. For example, if our sample included the Soviet Union, the expenditure of one country would far exceed that of any of the others and would therefore skew the distribution. The median would then be a more appropriate centrality measure, since it would divide the distribution in half. The Soviet Union would have no greater effect than any other nation on the location of the median.

Psychology

A good choice.   The mean number of points assigned will yield the best estimate of what the typical newborn infant is capable of doing (at least in your particular sample of infants). If, for example, the mean number of points assigned were 50, then 50 would be a standard or norm for judging the normality of a particular infant, and a child scoring far below 50 could be viewed as abnormally delayed.

Possible misinterpretations.   The neonates selected from the hospitals in your city may be different from neonates in general. For example, if your sample were composed mainly of Black infants, the mean might be too high to be representative of infants in general because Black infants tend to develop more rapidly than do those of other races.

Social Work

A good choice.   The mean can give you an estimate of the average number of hours of service families receive from this agency.

Possible misinterpretations.   A single family that requires a particularly intensive treatment program with many hours of service raises the mean substantially unless the N is very large. Even with a large N, a few really extreme cases can distort the mean as a measure of central tendency. In general, the mean should be used only when the distribution is reasonably symmetrical.

Sociology

A good choice.   The mean can give you an estimate of the average income of residents in the city.

Possible misinterpretations. The mean is sensitive to extreme scores. If the distribution includes such scores, and if they do not happen to be arranged so that they balance each other, the mean will give you misleading information about the bulk of the population.

Chapter 4

Education

A good choice.   The standard deviation can tell you whether the variability of the ratings for the two packages is the same. One package may be fairly uniformly rated by the great majority of teachers; that is, teachers may not differ very greatly in their assessments of the usefulness and practicality of the techniques presented in the package. The other training program may be rated very high on practicality and usefulness by some teachers and very low by others, resulting in a high standard deviation. The training program with the relatively large standard deviation may be less desirable because it may result in widely varying teacher percep-tions about the usefulness and practicality of the techniques, lead-ing to a less uniform adoption of the techniques than if the first package were used.

Possible misinterpretations.   Misinterpretations include attributing the higher variability of the second in-service training program entirely to problems inherent in the package. It may be due partly or even entirely to other factors—differences in the time of day the teachers were trained, for example, or differences in administrative support from one school to another.

Political Science

A good choice.   The standard deviation can tell you how much Latin American countries differ with regard to the incidence of military coups. If each country experiences approximately the same number of coups, there are almost no differences and the average deviation around the mean is nearly zero. But the more differences there are, and the larger they are, the larger is the average deviation from the mean. The standard deviation is a kind of average of the deviations of individual countries from their mean.

Possible misinterpretations.   As in the case of the mean, a few extreme scores within a frequency distribution can give misleading results when the standard deviation is calculated. Therefore, if only a few Latin American countries were to experience a very high number of coups, then just as the median would be a better measure of central tendency than the mean, so the semi-interquartile range (or some other quartile deviation) would be a better measure of variation than the standard deviation.

Psychology

A good choice.   The standard deviation provides a measure of consistency (agreement)—or rather, a measure of inconsistency (disagreement)—among the observers. If, for instance, some observers report as few as two aggressive acts per day and others report as many as twenty, the standard deviation will be large, and the reliability of the ratings should be questioned.

Possible misinterpretations.   Given a high level of variability among the five observers in our example, it would be tempting to blame one or several raters (e.g., the least experienced) for the apparent discrepancies and inaccuracies. A more likely explanation would be that the instructions to the observers failed to clarify what constitutes an aggressive act. If so, the instructions would need be revised.

Social Work

A good choice.   The standard deviation can tell you the amount of variability in grant awards to member agencies. A high standard deviation would indicate considerable differences in the amounts of money received by agencies. A low standard deviation would indicate that agencies were funded at approximately the same level.

Possible misinterpretations.   Again, as with the mean, extreme values can result in a higher standard deviation; thus, while only one or two agencies might be receiving much more or much less money than all the other agencies, a large standard deviation makes it appear that there is much variation throughout the distribution. In general, the standard deviation should be used only when the distribution is approximately normal.

Sociology

A good choice.   The standard deviation can give an estimate of that span of family sizes which includes 68 percent of all families and that which includes 96 percent of all families. By referring to normal curve tables, intermediate spans can be obtained if you know the standard deviation, and the regions above and below the mean can be examined separately.

Possible misinterpretations.   All these inferences are based on the assumption of a normal distribution of family sizes. If that distribution is not normal, you will have to find other ways of reporting your conclusions. (See the section on semi-interquartile range.)

Chapter 5

Education

A good choice.   The Pearson *r* can tell you whether there is a relationship between self-concept and social responsibility. The Pearson *r* will also tell you whether the relationship between self-concept and social responsibility is positive (+) or negative (−). The size of the coefficient, whether positive or negative, will indicate the strength of the relationship between the two measures.

Possible misinterpretations.   From a strong correlation (.70 or .80), it might be erroneously concluded that a student's self-concept causes him to be socially responsible or irresponsible. (Although the two variables do vary together, this may be due to some third variable, such as previous school or home experiences, which affects both social responsibility and self-concept.) It might also be erroneously concluded that a program effective in enhancing a person's self-concept would necessarily increase social responsibility. (If self-concept is caused by social responsibility, or if both are caused by some third factor or set of factors, changing self-concept will not affect social responsibility.)

Political Science

A good choice.   The Pearson *r* can tell you the magnitude and direction of the relation between the amount of conflict within countries and the amount of foreign conflict they initiate. First, the larger the coefficient, the stronger the association. Second, a positive sign indicates a direct relationship between *X* and *Y*, while a negative sign indicates an inverse relationship.

Possible misinterpretations.   Normally, a strong positive value of *r* would be taken as evidence that domestic conflict is associated with foreign policy conflict. Besides remembering that correlations do not necessarily entail causal relations, you should be aware of the assumptions behind *r*. If these assumptions are violated, then the results may not be reliable. The most important of these assumptions are that the relationship between *X* and *Y* is linear (straight regression line) and that the variance of the data points around that line is about the same everywhere along the line.

Psychology

A good choice.   The Pearson *r* will indicate the degree of association (.00 to either $+1.00$ or -1.00) between sugar levels and activity ratings for the 100 children. The relationship can be *positive* (high sugar levels associated with excessive activity and low sugar levels associated with low activity) or *negative* (high sugar levels associated with low activity and low sugar levels associated with high activity).

Possible misinterpretations.   A significant positive or negative relationship does not imply a cause-and-effect relationship. A positive Pearson *r* does not necessarily mean that high sugar levels cause hyperactivity. Hyperactivity may stem from a fundamental *impulsivity* in children. That is, the child's inability to control his impulses may lead to both excessive activity *and* excessive ingestion of sugar. Thus a third factor (impulsivity) may be responsible for an observed relationship between blood sugar and hyperactivity.

Social Work

A good choice.   The Pearson *r* provides a measure of association that reveals both the strength and the direction of the relationship between the two variables. Your hypothesis is that women of high self-esteem are less dependent on welfare than do women of low self-esteem. A high negative correlation would confirm your hypothesis.

Possible misinterpretations.   Even if your hypothesis is confirmed and you know self-esteem and requesting welfare are related, you don't know just *how* they are related. Two possibilities are (1) self-esteem renders a woman more determined to make herself economically self-sufficient, and (2) inability to make herself self-sufficient damages her self-esteem.

Sociology

A good choice.   The Pearson *r* can tell you whether there is a relationship between the conservativism of academic disciplines and the authoritarianism of their professors.

Possible misinterpretations.   Misinterpretations include the notion that the first variable causes the second or vice versa. In fact, both variables may be products of an unnamed external factor or set of factors.

Chapter 6

Education

A good choice.   If you transform raw scores for all the scales and subscales of the test into standard *z* scores based on performances by students of the same age and grade, it will be easier for you to compare the student's performance on the various subtests. For example, a *z* score of 0 would indicate that the student is average in that ability in comparison to his peers. A *z* score of $+1$ would indicate that the student is one standard deviation above the mean (at the 84th percentile) in that ability in comparison to fourth-grade students nationwide. A *z* score of -2 would indicate that he is two standard deviations below the mean (at the 2nd percentile) of his grade. Such comparisons can help the teacher to determine more exactly in what areas the student is having difficulty so that some specific assignments can be designed to strengthen those abilities.

Possible misinterpretations. Misinterpretations include regarding ability scores as direct indices of hereditary potential. For instance, if a student is a member of a lower-class ethnic minority and the test has been normed on a middle-class white population, his relatively low abilities may be due to environmental rather than hereditary limitations.

Political Science

A good choice. The z score can tell you where each score fits on a scale common to all. The z scale is based on each distributions mean and standard deviation. In this case, each of the various civil strife scores is expressed in terms of the standard deviation of its own distribution and measured from its own mean. Once all the scores have been converted to z scores, every distribution has the same mean (0) and standard deviation(1). *Now* it makes sense to combine them.

Possible misinterpretations. Don't regard standard scores as absolute in the sense that physical measures (e.g., length) are absolute. Each score tells only where a given measure lies within a distribution of measures of the same attribute. In a different distribution, the same measure might yield a different standard score.

Psychology

A good choice. By transforming each raw score to a new scale with a mean of 0 and a standard deviation of 1, you can ascertain whether June is developing normally in all three domains. If she earns z scores of -1 on all three tests, you can infer generally delayed development. If all three z scores are $+1$, you can infer advanced development. If she obtains z scores of $+2$ (intelligence), -1 (social), and -2 (psychomotor), you can infer fragmented development.

Possible misinterpretations. One possible misinterpretation is the presumption that obtained measurements are stable across time. Actually, the behavior of infants and young children tends to vary a great deal from one observation to the next.

Social Work

A good choice.   By converting each raw score into a standard score with a mean of 0 and a standard deviation of 1, you place all scores on a standard scale, thus making it possible to compare them.

Possible misinterpretations.   If the various tests have been standardized on different populations, comparisons will be risky. For example, scoring high on client assessment in a population of beginners may not be a more laudable accomplishment than scoring low on client treatment in a population of highly trained professionals.

Sociology

A good choice.   The z score can tell you how this professor compares to other professors in your sample in terms of standard deviation units above or below the mean.

Possible misinterpretations.   The sample is not a true representative of the population of interest. In this case, the sample was taken from a population defined as *all* professors at this university. Professors teaching the courses for which you are eligible may differ in some significant way from the total faculty (e.g., they may have been selected for their ability to relate to lower-division students). If so, the English professors's z score may well lead you to make a bad decision at registration time. (For example, his z score on authoritarianism could be low in the general faculty but high among those faculty who are teaching these particular courses.)

Chapter 7

Education

A good choice.   The standard error of the mean is used to establish a confidence interval within which lies the population mean for each grade level. The band of scores described by the confidence interval provides a better index than the obtained mean itself of the achievement level of each local grade because it takes into account the errors that inevitably occur whenever measurements are made. You can say that the probability is _____ percent that the mean "really" is some value between _____ and _____ .

Possible misinterpretations. Don't assume that if the band provided by the confidence interval around a sample mean is higher or lower than the national mean, the difference is due solely to the school program. Other factors such as the influences of the home and the community must be considered also.

Political Science

A good choice. Precisely these limits are described by the *confidence interval.* There is a confidence interval for any desired probability. For example, say you want to specify the interval within which you can place the true mean at a confidence level of 68 percent (i.e., the interval for which the probability of the true mean's location within it is 68 percent). Your obtained mean is 50. If the standard error ($SE_{\bar{x}}$) is 10, the interval is $2 \times 10 = 20$ points wide. (See pages 82–85.) At the $68X$ percent level of probability, then, the confidence interval would extend from 40 to 60 points.

Possible misinterpretations. You might think it equally probable for the true mean to be at any point within the confidence interval. Actually, the probability is much higher in the middle of the interval than anywhere else. Your obtained mean is, after all, your best estimate of the true mean.

Psychology

A good choice. The confidence interval would allow you to estimate the *limits* within which the boy's true inkblot test score resides. It would also provide a statement of the *probability* that his true score is indeed within those limits. If the limits were far apart or the probability low, you would be cautious in your interpretation of the boy's score.

Incidentally, the manual might include, either in place of the confidence interval or in addition to it, a correlation coefficient (r_{xx}). This statistic represents another way of looking at reliability. (See Chapter 5.)

Possible misinterpretations.   You may be tempted to think of "error" as something that *lowers* a score. But measurement error (unreliability) can also raise a score. The confidence interval therefore extends both *below* and *above* the obtained score. (The obtained score is in the middle of the interval.)

Social Work

A good choice.   The confidence interval provides a range of scores within which the family's true FLIP rating probably resides. There are two ways to approach the question "Where is the true score?"

You can (1) set limits above and below the obtained score and then find the probability that the *true* score is between those limits; or (2) set an acceptable probability and then set the limits that match it. In either case, the interval between the limits is called the *confidence interval,* and the probability is a *level of confidence.*

Possible misinterpretations.   In tests of significance (Chapter 8) the probability cited is that of an occurrence *outside* prescribed limits. Here, you report the probability that the true mean is *inside* those limits.

Sociology

A good choice.   The standard error of the mean can give you a band of scores (the confidence interval) that probably contains the true mean. It can also specify the level of that probability (the level of confidence).

Possible misinterpretations.   The universal mean does not *have* to fall inside the intervals you have computed—i.e., don't accept the specification of a confidence interval without mentioning the corresponding level of confidence.

Chapter 8

Education

A good choice.   The *t* ratio can tell you whether at the end of the semester the students who were in the group counseling program differ significantly in teacher-rated disruptiveness from the students who were not in the program. Thus the staff could determine (with a given probability) whether students in the group counseling program are rated as less disruptive at the end of the semester than are the students not in the program.

Possible misinterpretations.   You can't necessarily attribute differences between the groups solely to the counseling program. Other factors may have influenced the results:

1. Even though the groups were randomly assigned, there may have been initial differences in disruptiveness at the beginning of the semester.
2. The teachers knew which students were in the treatment program, and that may have influenced them to react to those students differently or to rate them differently even though they actually are not different from students who are not in the counseling program.
3. It may be that just the added attention of the school counselor or just the inclusion in a new program would have resulted in a lowering of disruptiveness regardless of the content of the program.

Political Science

A good choice.   The *t* test can be used to determine whether the observed difference between the means of the two groups is due to chance. The average difference between the two means within repeated pairs of random samples taken from the same population would be zero. If the observed difference is large enough to be statistically significant, then we can reject the hypothesis that the difference is merely a chance difference between two random samples of the same population of crime-prevention efforts. We conclude rather that the new citizen-participation program is truly different from the old cops-only program with respect to the prevention of burglary.

Possible misinterpretations.   While the *t* test is appropriate for analyzing the difference of means for two small samples from the same or identical populations, its use is subject to three restrictions. First, the observations in the two samples must be independent of each other. Second, the populations must not be skewed in opposite directions. Finally, if the populations do not have equal variances, adjustments are needed in the way in which *t* is calculated.

Psychology

A good choice.   The *t* ratio can tell you whether the relaxation treatment or the drug therapy is the more effective in terms of the mean scores of the two groups. It will also give you the probability that there really is no difference between them—that they are both random samples from a single population with respect to activity level.

Possible misinterpretations.   Although the results may show a significant difference between the two groups, locating the *cause* of that difference may be difficult. Suppose that the drug therapy group appears to benefit more from treatment. That could be the result of the drug, but unless you have equalized the age of the two groups, age differences could be the critical factor. Without controlling alternative causes, you cannot know whether the cause is the drug or some other factor such as the ages of the children, their intelligence, or characteristics of their drug therapists.

Social Work

A good choice.   The *t* ratio can be used to determine whether there is a significant difference (difference not due to chance alone) between two groups. The difference is significant if you can reject the null hypothesis that the two groups are from the same population with respect to health. You can take your *t* ratio to a table that will give you the probability that the two groups *are* from the same population. If that probability is extremely small—say, .01—you may assert with considerable confidence that the experimental program has been effective.

Possible misinterpretations.   Your confidence in the program will have been misplaced if some outside factor influences one group but not the other. For example, if the experimental group rides a special bus to the center, eats its meals together, or does something else together as a by-product of the program, it could be all that "togetherness" that improves the seniors' health, rather than the program itself.

In confidence intervals (Chapter 7), the probability cited is *inside* prescribed limits. Here, you report the probability that the true mean is *outside* those limits. The .01 cited above, for example, indicates that by chance 99 percent of differences between groups taken at random from a single population would be inside the prescribed limits. Only 1 percent would extend beyond them.

Sociology

A good choice.   The *t* ratio, entered into appropriate tables, can tell you how likely it is that a difference this large would occur by chance.

Possible misinterpretations.   As in correlation, a relationship does not by itself justify an inference of causality. If most of the Catholics in your state belong to a different social class than most non-Catholics, it might not be legitimate to infer that the difference in religious belief is the cause of the difference in family size. It might turn out that with social class held constant, Catholic families are no larger than non-Catholic.

Chapter 9

Education

1. *A good choice.*   Chi square can tell you whether significantly more students from the vocational training program graduate than are expected from your knowledge of the number of students who graduate from the group of at-risk students who applied but were not accepted into the vocational training program.

Possible misinterpretations.   Don't conclude that the training program is the only cause of the observed difference. Even if the program significantly increases the percentage of at-risk students who graduate, its success may be due more to the interest of the staff in a new program than to any attribute of the program *per se.*

2. *A good choice.*   One-way analysis of variance can tell you whether the difference between any two or more of the three

social-problem-solving test means is greater than would be expected by chance. If there is such a difference (and it is in favor of the clients of the trained counselors), then you may conclude that the students who went through the program are better able to solve the problems posed by the test than are students in the other two groups.

Possible misinterpretations.   Don't assume that if there is a significant F ratio, each group mean differs significantly from each of the others. For the F test to be significant, it is necessary only that two of the means differ significantly. You may also erroneously conclude that the test result must be attributable to the program. Even though the students were randomly assigned, there could have been differences between the groups before the programs began. The students could have learned which groups they were in after the study began and been affected by that knowledge. Similarly, each counselor's knowledge of his place in the program might affect student behavior in ways not specified by the training he received.

3. *A good choice.*   Two-way analysis of variance of the abstract-reasoning-ability scores from the beginning of the year can tell you whether there is any difference among the mean test scores of the students assigned to any two or more of the programs before they begin. The same analysis can tell you whether there is a significant difference in reasoning ability between the sixth and seventh grades prior to your intervention. The analysis can tell you also whether, before the programs begin, there are significant interactions between (1) abstract reasoning ability in the four programs and (2) grade level.

At the end of the year, a two-way analysis of variance of the abstract-reasoning-ability test can tell you whether there is a difference between two or more of the groups that is not attributable to chance. Finally, you can tell whether there is a significant interaction between type of program and grade level at the end of the year.

Possible misinterpretations.   Attributing significant differences at the end of the year to the effectiveness of one or more programs would be an error if the difference was actually there in the beginning.

Political Science

1. *A good choice.*   Chi square can tell you the probability that any deviation of the observed frequencies from a stipulated expected frequency is due to chance. (The null hypothesis here is that equal proportions of Republicans and Democrats support Salt II.) Chi square compares the observed frequencies in each cell of the contingency table with what would be expected in these cells if the two variables—party affiliation and support for Salt II—were independent.

Possible misinterpretations.   Misuse of chi square usually occurs when certain key assumptions are violated. The most important of these is the assumption that the data represent a random sample of independent observations. In this case, you must be sure that you sample *all* Republicans, not just some subgroup dedicated to the defeat of the treaty, and that your sample of Democrats is similarly random.

2. *A good choice.*   One-way analysis of variance can tell you whether observed differences (in coup frequency) among types of regime are likely to have occurred by chance. The analysis of variance is an extension of the difference-of-means t test which we examined earlier. Had there been only two types of regime legitimacy, a t test would have sufficed. With more than two types, however, you must compute an F ratio. If the population variance in coup frequency estimated from differences between the countries is not significantly greater than the variance as estimated from within the groups, then you cannot reject the null hypothesis. That is, you must conclude that the differences you have observed are merely random variations—that the three types of country are all one with respect to frequency of coups.

Possible misinterpretations.   When the requirements of the one-way analysis of variance are met, the F ratio allows you to determine whether the observed difference among the group means is large enough to be statistically significant. However, the F ratio does not tell you which independent variables are associated with what observed differences. Further tests are necessary to accomplish that. (See Appendix I, Multimean Comparison Tests.)

3. *A good choice.* Two-way analysis of variance can tell you whether there is a significant "form of government main effect" and whether there is a significant "type of leadership main effect." It can also tell you whether there are any "interaction effects." For example, it might turn out that totalitarian governments are more aggressive than other forms only when their leadership is of the unitary type.

Possible misinterpretations. The two-way analysis of variance is subject to the same kinds of limitations as the one-way analysis of variance. Specifically, the F test does not tell you precisely where among the nine categories the significant differences lie. Again, further tests can be made. (See Appendix I, Multimean Comparison Tests.)

Psychology

1. *A good choice.* Chi square can tell you whether the number of children who pass the test after treatment is greater or smaller than the number that might be expected to pass without treatment. If previous research indicates that 45 percent of untreated agoraphobics recover spontaneously, this figure might be used as the expected value (the null hypothesis).

Possible misinterpretations. In this design, uncontrolled factors could possibly account for significant findings. For example, just coming to the clinic may be sufficient to induce change, or those who come to the clinic may not be a random sample of agoraphobic children generally.

2. *A good choice.* One-way analysis of variance can tell you whether there are significant differences among the three treatments in their self-ratings of experienced pain.

Possible misinterpretations. You might infer from the F ratio that every mean differs from every other, but a significant F ratio tells you only that there is a difference *somewhere* between or among treatment means. You now have to scrutinize your data to ascertain where the difference(s) is(are). (See Appendix I.)

3. *A good choice.* A two-way analysis of variance allows you to test: (1) whether any of the four group means (introvert-

individual, introvert-group, extravert-individual, extravert-group)
is significantly different from any other; (2) whether group therapy
or individual therapy is the more effective, regardless of introversion-extraversion tendencies; (3) whether introversion-extraversion tendencies are associated with better outcome, regardless of
the type of therapy; and (4) whether there is an interaction
between introversion-extraversion and therapeutic modality—i.e.,
whether individual therapy works best with introverts, group therapy with extraverts, as you had predicted.

Possible misinterpretations. In this particular example,
there are only two categories of each variable (introvert-extravert,
individual-group). Whenever that is true, the F ratio for each variable identifies precisely the source of any difference that we find.
For example, if the F for the introvert-extravert variable turned out
to be significant, we would know that the difference is between
the group we have designated "introvert" and the one that we call
"extravert."

When there are more than two categories of a variable, however, the F test does *not* identify the source precisely. If, for example, we had divided our introvert-extravert variable into three categories (introvert, ambivert, extravert) instead of two, a significant
F would mean only that there was a difference somewhere within
that variable. It would not tell us whether that difference was (1)
between introverts and ambiverts, (2) between introverts and
extraverts, or (3) between ambiverts and extraverts. Further tests
would be necessary for more precise identification. (See Appendix
I, Multimean Comparison Tests.)

Social Work

1. *A good choice.* Chi square can tell you whether the frequency of recidivism is significantly lower in the group of youth
appearing before the Board than in the group who are sent to court.
Chi square is an index of deviation from the frequencies you would
expect if the new Board's procedures were not any more (or less)
effective than the courts'. Indeed, chi square *is* that deviation,
expressed as a proportion of the expected frequencies. In the tables
above, the *expected* and *obtained* frequencies are displayed along
with the *differences* between the two. There are 70 recidivists and
130 non-recidivists. Since the Board and the court groups contain

equal numbers of juveniles, the null hypothesis is that half of the 70 recidivists and half of the 130 non-recidivists are in each group. Half of 70 is 35 and half of 130 is 65; those are the expected frequencies. Once the data are recorded, the first question is, "Do the obtained frequencies differ from the expected?" It is clear from the "difference" table that they do. (The signs of the differences have been omitted. They are of no concern to us, and they are in fact eliminated in the process of computing Chi Square.) The next question, however, is, "Are the obtained differences statistically significant?" (What is the probability that differences this large, taken together, would occur by chance?) To answer *that* question, you must compute chi square and take it to the appropriate probability table (in a book other than this one).

Recidivists

	Yes	No
Board	35	65
Court	35	65

Expected

Recidivists

	Yes	No	
Board	30	70	100
Court	40	60	100
	70	130	

Obtained

Recidivists

	Yes	No
Board	5	5
Court	5	5

Differences

Possible misinterpretations. Chi square deals with frequency counts only. Every score is therefore required to be either 0 or 1, and information about the *magnitude* of offenses is lost. It is even possible that the relatively few offenders who break the law in spite of the Board's rehabilitation efforts are guilty of offenses more serious than those committed in larger numbers by the control group.

2. *A good choice.* One-way analysis of variance can tell you whether there is a significant difference anywhere among the four groups.

Possible misinterpretations. A significant F does *not* mean that every group is different from every other with respect to hyperactivity. It tells you only that there is a difference *somewhere* among the groups. If the F test does reveal a significant difference, you can then follow up with other tests especially designed for use after the F test. Those tests will locate the difference(s). (See Appendix I, Multimean Comparison Tests.)

3. *A good choice.* A two-way analysis of variance can tell you whether there are significant differences among the four group means, i.e., 1) children, foster family; 2) youth, foster family; 3) children, group home; and 4) youth, group home. Two-way analysis of variance also identifies the *location* of the effect(s)—i.e., whether adjustment is affected by age, by type of foster care, or by their interaction.

Possible misinterpretations. Two-way analysis of variance yields an F ratio for each main effect and the interaction, so it is not subject to the error described earlier for one-way analysis of variance. The location of any difference(s) is (are) known from the first computation.

Sociology

1. *A good choice.* Chi square can tell you whether or not you could expect the numbers in the cells of your 2×2 table (early vs. late $\times$ approval vs. disapproval) to have occurred the way they did on the basis of chance alone. The other possibility is that these

numbers were influenced by something other than chance; in this case, the belief about when human life begins is a good candidate.

Possible misinterpretations. Possible misinterpretations include regarding the chi square as a general—though probabilistic—answer to the question implied by your hypothesis. The question is, "Does a person's belief about when human life begins affect his attitude toward abortion?" You might get a different answer if, for example, "early" and "late" were defined in relation to nine months instead of ninety days. Or your respondents' attitudes toward abortion might look different if the question were phrased ". . . under any circumstances?" instead of ". . . on demand?", and that, too, might change the answer to the general question.

2. *A good choice.* One-way analysis of variance will give you the probability that all of the obtained differences occurred by chance. If that probability is very small, at least one of the differences is statistically significant.

Possible misinterpretations. A significant F does not mean that each sample is significantly different from every other. The F test indicates only that there is at least one significant difference among those identified. If the F test is positive, then each mean must be compared with each other mean, and each difference must be evaluated separately from the others. (See Appendix I.)

3. *A good choice.* Two-way analysis of variance is an appropriate method. The summary table of F ratios will tell you whether there is a main effect of church attendance, whether there is a main effect of education, and whether there is an interaction between the two.

Possible misinterpretations. In this particular example, there are only two categories of each variable (church attendance–nonattendance, high education–low education). Whenever that is true, the F ratio for each variable identifies precisely the source of any difference that we find. For example, if the F for the attendance-nonattendance variable turned out to be significant, we would know that the difference was between the group we have designated *Church Attendance* and the one we call *Nonattendance.*

When there are more than two categories of a variable, however, the F test does *not* identify the source precisely. If, for example, we had divided our attendance–nonattendance variable into three categories (high, medium, and low attendance) instead of two, a significant F would mean only that there was a difference somewhere within that variable. It would not tell us whether that difference was (1) between high and medium, (2) between high and low, or (3) between medium and low attendance. Further tests would be necessary for more precise identification. (See Appendix I, Multimean Comparison Tests.)

SUGGESTED READINGS

Many valuable texts discuss the application of statistical principles to problems in business, economics, education, psychology, sociology, and other related fields. Here is a very small sample of what is available:

Alder, H. L., and E. B. Roessler. *Introduction to Probability and Statistics*, 6th ed. San Francisco: W. H. Freeman and Company, 1977.

Champion, D. J. *Basic Statistics for Social Research*. New York: Harper and Row, 1970.

Dornbusch, S. M., and C. F. Schmid. *A Primer of Social Statistics*. New York: McGraw-Hill, 1955.

Downie, N. M., and R. W. Heath. *Basic Statistical Methods*. 4th ed. New York: Harper and Row, 1974.

Freeman, L. C. *Elementary Applied Statistics: For Students in Behavioral Science*. New York: John Wiley and Sons, 1965.

Games, P., and G. Klare. *Elementary Statistics: Data Analysis for the Behavioral Sciences*. New York: McGraw-Hill, 1967.

Glass, G., and J. Stanley. *Statistical Methods in Education and Psychology*. Englewood Cliffs, N.J.: Prentice-Hall, 1970.

Guilford, J. P., and B. Fruchter. *Fundamental Statistics in Psychology and Education*, 5th ed. New York: McGraw-Hill, 1973.

Klugh, H. *Statistics: The Essentials for Research*, 2nd ed. New York: John Wiley and Sons, 1974.

Mason, R. D. *Statistical Techniques in Business and Economics*. 4th ed. Homewood, Ill.: Richard D. Irwin, 1978.

Meyers, L. S., and N. E. Grossen. *Behavioral Research: Theory, Procedure, and Design.* 2nd ed. San Francisco: W. H. Freeman and Company, 1978.

Moore, D. S. *Statistics: Concepts and Controversies.* San Francisco: W. H. Freeman and Company, 1979.

Shao, S. P. *Statistics for Business and Economics.* 2nd ed. Columbus, Ohio: Merrill, 1972.

Willemsen, E. W. *Understanding Statistical Reasoning: How to Evaluate Research Literature in the Behavioral Sciences.* San Francisco: W. H. Freeman and Company, 1974.

INDEX

DATE DUE

MAR 17 1999	